ORIENTED PROJECTIVE GEOMETRY

A Framework for Geometric Computations

JORGE STOLFI

Digital Systems Research Center
Palo Alto, California

ACADEMIC PRESS, INC.
Harcourt Brace Jovanovich, Publishers

Boston   San Diego   New York
London   Sydney   Tokyo   Toronto

ACADEMIC PRESS, INC.
1250 Sixth Avenue, San Diego, CA   92101

United Kingdom Edition published by
ACADEMIC PRESS LIMITED
24–28 Oval Road, London NW1 7DX

Library of Congress Cataloging-in-Publication Data
Stolfi, Jorge.
    Oriented projective geometry : a framework for geometric
computations / Jorge Stolfi.
          p.   cm.
    Includes bibliographical references and index.
    ISBN 0-12-672025-8 (alk. paper)
    1.Geometry. Projective — Data processing.   I. Title.
QA471.S88 1991
516'.5—dc20   91-16219
                                  CIP

Printed in the United States of America
91 92 93 94   9 8 7 6 5 4 3 2 1

TABLE OF CONTENTS

Introduction

Programmers who use homogeneous coordinates for geometric computations are implicitly — and often unknowingly — working in the so-called *projective space*, a strange and wonderful world which only superficially resembles the Euclidean space we all know and love. One difference is that certain concepts that are fundamental to geometric computing, such as segments, triangle orientation, and convexity, cannot be consistently defined in the projective world.

Oriented projective geometry is an alternative geometric model that combines the elegance and efficiency of projective geometry with the consistent handling of oriented lines and planes, signed angles, segments, convex sets, and many other concepts that the classical theory does not support. In this monograph I advance the thesis that *oriented* projective space — especially in its analytic form, based on *signed* homogeneous coordinates — is a better framework for geometric computations than their classical counterparts.

The differences between the classical and oriented models are largely confined to the mathematical formalism and its interpretation. Computationally, the differences are minimal; most geometric algorithms that use homogeneous coordinates can be easily converted to the oriented model with negligible effect on their performance. For many algorithms, the required changes are largely a matter of paying attention to the order of operands and to the signs of coordinates.

It is not the aim of this monograph to push the remote frontiers of mathematics or computer science. Theoreticians will not find here any deep theorems, intricate algorithms, or sophisticated data structures. Expert geometers will notice that oriented projective geometry is just anothr name for spherical (or double elliptic) geometry, which to them is an old and well-explored subject.

On the other hand, graphics programmers may be surprised to learn that the curved surface of the sphere is an excellent model for computations dealing with straight lines on the flat Euclidean plane. The aim of this monograph is to point out the value of this model for practical computing, and to develop it into a rich, consistent, and effective tool that those programmers can use in their everyday work. In keeping with this goal, I have strived to keep formal derivations and mathematical jargon to a minimum, and (at the risk of being tedious) to illustrate many general definitions and theorems with explicit examples in one, two, and three dimensions.

Here is a brief outline of this book. Chapter 1 gives a quick overview of classical and oriented projective geometry on the plane, and discusses their advantages and disadvantages as computational models. Chapters 2 through 7 define the canonical oriented projective spaces of arbitrary dimension, the operations of join and meet, and the concept of relative orientation, and study their properties. Chapter 8 defines projective maps, the space transformations that preserve incidence and orientation; these maps are used in chapter 9 to define abstract oriented projective spaces. Chapter 10 introduces the valuable notion of projective duality. Chapters 11, 12, and 13 deal with additional concepts related to projective maps, namely projective functions, projective frames, relative coordinates, and cross-ratio. Chapter 14 tells about convexity in oriented projective spaces. Chapters 15, 16, and 17 show how the affine, Euclidean, and linear vector spaces can be emulated with the oriented projective space. Finally, chapters 18 through 20 discuss the computer representation and manipulation of lines, planes, and other subspaces.

This monograph is a slightly edited and revised version of my Ph.D. dissertation, *Primitives for Computational Geometry*, submitted to Stanford University in May 1988, and printed under the same title as Technical Report 36 of the DEC Systems Research Center in January 1989. The present edition incorporates innumerable small corrections and improvements over these earlier versions.

Acknowledgements: I would like to express here my gratitude, first of all, to my advisor Leo Guibas, who helped me, prodded me, and supported me in more ways that I could possibly list here; and to my boss Bob Taylor, who gave me constant encouragement and the opportunity to experience the unique research environment that he created at Xerox PARC and DEC SRC.

I learned a great many things about this subject from discussions with my colleagues, especially with Lyle Ramshaw, John Hershberger, Mike Lowry, and Yuen Yu. I am grateful to Professors Ernst Mayr and Andrew Yao for their careful reading of my dissertation. From Mary-Claire van Leunen and Cynthia Hibbard I received a good deal of advice on the art and science of writing, which immensely improved my syntax and style — from truly atrocious to, I hope, merely dreadful.

I am also indebted to Ken Brooks, Marc Brown, Bill Kalsow and Lyle Ramshaw for taking time out of their own research to maintain the software on which I depended; and to the Digital Equipment Corporation, the Xerox Corporation, the University of São Paulo, and the Conselho Nacional de Desenvolvimento Científico e Tecnológico (CNPq) of Brazil for their generous financial support.

Finally, I wish to give my deepest thanks to all my dear friends, here and abroad, who never grew tired of asking when would I be finished; and to my wife Rumiko, who deserves more credit for this work than all of the above put together.

Jorge Stolfi
March 1991

Chapter 1
Projective geometry

The bulk of this chapter is a quick overview of the standard (unsigned) homogeneous coordinates for the plane, and the classical (unoriented) projective geometry which they implicitly define. In order to motivate what follows, I will discuss at some length the advantages and disadvantages of homogeneous coordinates as a computational model, compared to ordinary Cartesian coordinates. The chapter concludes with a quick overview of *oriented* projective geometry, the alternative computational model which is the subject of this book, and which I define formally in the following chapters.

The description of projective geometry given below below is necessarily sketchy, and does not even begin to make justice to the richness and elegance of the subject. Mathematically inclined readers who wish to know more are urged to start from any basic textbook on the subject, such as Coxeter's [6], and follow the leads from there. Readers interested in practical applications of projective geometry to computer graphics are advised to read the the book by Penna and Patterson [16].

1. The classic projective plane

The projective plane can be defined either by means of a "concrete" model, borrowing concepts from linear algebra or Euclidean geometry [15], or as an abstract structure satisfying certain axioms [4, 6].

Definitions that follow the axiomatic approach have the advantage of being concise and elegant, but unfortunately they cannot be generalized easily to spaces of arbitrary dimension. Moreover, the axiomatic approach seems better suited to formalizing intuitive knowledge already acquired, than at developing and teaching such knowledge. Therefore, considering the aims of this monograph, I have chosen to avoid the axiomatic approach, and to base all definitions on four concrete models of projective space: the *straight*, *spherical*, *analytic*, and *vector space* models, which are described below.

1.1. The straight model

The *straight model* of the projective plane $\mathbf{P}_2$ consists of the real plane $\mathbf{R}^2$, augmented by a *line at infinity* Ω, and by an *infinity point* $d\infty$ for each pair of opposite directions $\{d, -d\}$. The point $d\infty = (-d)\infty$ is by definition on the line Ω and also on every line that is parallel to the direction d. See figure 1.

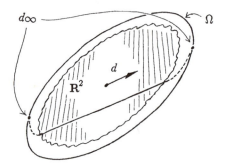

Figure 1. The straight model of the projective plane $\mathbf{P}_2$.

1.2. The spherical model

The *spherical model* of $\mathbf{P}_2$ consists of the surface of a sphere, with diametrically opposite points identified. The lines of $\mathbf{P}_2$ are represented by the great circles of the sphere, again with opposite points identified. See figure 2.

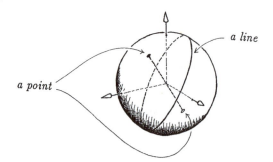

Figure 2. The spherical model.

The spherical model clearly shows that all lines and points are equivalent in their topological and incidence properties. The special role that Ω and the infinite points seem to play in the straight model is a mere artifact of the latter's representation.

1.3. The analytic model

The *analytic model* represents points of $\mathbf{P}_2$ by their *homogeneous coordinates*, and lines by their *homogeneous coefficients*. A point is by definition a non-zero triplet of real numbers $[w, x, y]$, with scalar multiples identified. In other words, $[w, x, y]$ and $[\lambda w, \lambda x, \lambda y]$ are the same point, for all $\lambda \neq 0$. A line is also represented by a non-zero real triplet $\langle W, X, Y \rangle$, which by definition is incident to all points $[w, x, y]$ such that $Ww + Xx + Yy = 0$. Note that $\langle W, X, Y \rangle$ and $\langle \lambda X, \lambda Y, \lambda Z \rangle$ are the same line for all $\lambda \neq 0$.

1.4. The vector space model

Geometrically, we can identify the point $[w, x, y]$ of $\mathbf{P}_2$ with the line of $\mathbf{R}^3$ passing through the origin and through the point (w, x, y). The line $\langle W, X, Y \rangle$ of $\mathbf{P}_2$ then corresponds to the plane of $\mathbf{R}^3$ passing through the origin and perpendicular to the vector (W, X, Y). In other words, we can identify points and lines of $\mathbf{P}_2$ with one- and two-dimensional linear subspaces of the three-dimensional vector space $\mathbf{R}^3$. In this way we get the *vector space model* of $\mathbf{P}_2$. See figure 3.

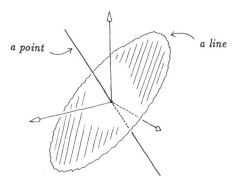

a point

a line

Figure 3. The vector space model of $\mathbf{P}_2$.

1.5. Correspondence between the models

The analytic and straight models of $\mathbf{P}_2$ are connected by the homogeneous-to-Cartesian coordinate transformation well-known to graphics programmers, which takes the homogeneous triplet $[w, x, y]$ is mapped to the point $(x/w, y/w)$ of the Cartesian plane. We can view this transformation as choosing among all equivalent homogeneous triplets a *weight-normalized* representative $(1, x/w, y/w)$ (the first co-ordinate w being called the *weight* of the triplet). As a special case, homogeneous triplets with weight $w = 0$ are mapped to the infinity points of the straight model.

The analytic and spherical models are connected by the transformation that takes the homogeneous triplet $[w, x, y]$ to the point

$$\frac{(w, x, y)}{\sqrt{w^2 + x^2 + y^2}}$$

on the unit sphere of $\mathbf{R}^3$.

Geometrically, these mappings corresponds to *central projection* of $\mathbf{R}^3$ onto the unit sphere, or onto the plane π tangent to the sphere at $(1, 0, 0)$. See figure 4. This projection takes a pair of diametrically opposite points p, p' of the sphere to the point q where the line pp' meets the tangent plane π. The great circle of the sphere that is parallel to the plane π is by definition projected onto the line at infinity Ω of the straight model. Observe how this correspondence preserves points, lines, and their incidence relationships.

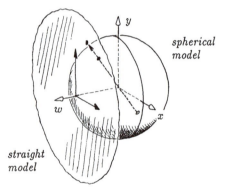

Figure 4. Central projection between the models of $\mathbf{P}_2$.

2. Advantages of projective geometry

Projective geometry and homogeneous coordinates have many well-known advantages over their Cartesian counterparts. From the point of geometrical computing, the following ones are particularly important:

- *Simpler formulas.* Projective geometry and homogeneous coordinates have many well-known advantages over their Cartesian counterparts. For one thing, the use of homogeneous coordinates generally leads to simpler formulas that involve only the basic operations of linear algebra: determinants, dot and cross products, matrix multiplications, and the like. All Euclidean and affine transformations, and all perspective projections, can be expressed as linear maps acting on the homogeneous coordinates of points. For example, the Cartesian coordinates of the point where the lines $ax + by + c = 0$ and $rx + sy + t = 0$ intersect are

$$\frac{(bt - cs, \ cr - at)}{as - br}$$

In homogeneous coordinates, the intersection of $\langle a, b, c \rangle$ and $\langle r, s, t \rangle$ is

$$[\, bt - cs, \ cr - at, \ as - br \,]$$

which is easily recognized as the cross product of the vectors (a, b, c) and (r, s, t). As this example shows, with homogeneous coordinates we can eliminate most of the division steps in geometric formulas; the savings are usually enough to offset the cost of handling an extra coordinate. The absence of division steps also makes it possible to do exact geometric computations with all-integer arithmetic.

- *Less special cases.* Homogeneous coordinates let us represent points and lines at infinity in a natural way, without any *ad hoc* flags and conditional statements. Such objects are valid inputs in many geometric applications, and are generally useful as "sentinels" in algorithms (in sorting, merging, list traversal, and so forth). They also allow us to reduce the number of special cases in theorems and computations. For example, when computing the intersection of two lines we don't have to check whether they are parallel. The general line intersection formula will work even in this case, producing a point at infinity. This point can be used in further computations as if it were any ordinary point. By contrast, in the Euclidean or Cartesian models we must disallow this special case, or explicitly test for it and handle it separately. Note that when we compose two procedures or theorems, the number of special cases usually grows multiplicatively rather than additively. Therefore, even a small reduction in the special cases of basic operations — say, from three to two — will greatly simplify many geometric algorithms.

- *Unification and extension of concepts.* Another advantage of projective geometry is its ability to unify seemingly disparate concepts. For example, the differences between circles, ellipses, parabolas, and hyperbolas all but disappear in projective geometry, where they become instances of the same curve, the non-degenerate conic.

  As another example, all Euclidean and affine transformations — translations, rotations, similarities, and so on — are unified in the concept of *projective map*, a function of points to points and lines to lines that preserves incidence. As is often the case with new unifying concepts, the class of projective maps turns out to include new interesting transformations, such as the perspective projections, that were not in any of the original classes. In fact, these maps cannot be properly defined in Euclidean geometry, since they exchange some finite points with infinite ones.

- *Duality.* Consider the one-to-one function '$*$' that associates the point $[w, x, y]$ to the line $\langle w, x, y \rangle$, and vice-versa. This mapping preserves incidence: if point p is on line l, then line p^* passes through point l^*. The existence of such a map ultimately implies that every definition, theorem, or algorithm of projective geometry has a *dual*, obtained by exchanging the word "point" with the word "line," and any previously defined concepts by their duals. For example, the assertion "there is a unique line incident to any two distinct points" dualizes to "there is a unique point incident to any two distinct lines."

  This *projective duality* is an extremely useful tool, in theory and in practice. Thanks to it, every proof automatically establishes the correctness of two very different theorems, and every geometrical algorithm automatically solves two very different problems. It turns out that a geometric duality with these properties can be defined only in the full projective plane. In the Euclidean plane one can construct only imperfect dualities, that do not apply to certain lines and/or points. The use of such pseudo-dualities often leads to unnecessarily complicated algorithms and proofs, with many spurious special cases [17].

3. Drawbacks of classical projective geometry

In spite of its advantages, the projective plane has a few peculiar features that are rather annoying from the viewpoint of computational geometry. Some of those problems, which were described in detail by Riesenfeld [19], are:

- *The projective plane is not orientable.* Informally, this means there is no way of defining "clockwise" or "counterclockwise" turns that is consistent over the whole plane $\mathbf{P}_2$. The reason is that a turn can be continuously transported over the projective plane in such a way that it comes back to its original position but with its sense reversed. For the same reason, it is impossible to tell whether two triangles (ordered triplets of points) have the same or opposite handedness. This limitation is quite annoying, since these two tests are the building blocks of many geometric algorithms.

- *Lines have only one side.* If we remove a straight line from the projective plane, what remains is a *single* connected set of points, topologically equivalent to a disk. Therefore, we cannot meaningfully ask whether two points are on the same side of a given line. More generally, *Jordan's theorem is not true* in the projective plane, since a simple closed curve (of which a straight line is a special case) need not divide the plane in two distinct regions. Even if we consider only the immediate neighborhood of a line, we still cannot distinguish its two sides, since that neighborhood has the topology of a Möbius band. See figure 5.

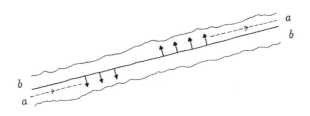

Figure 5. The neighborhood of a straight line of $\mathbf{P}_2$.

- *Segments are ambiguous.* In projective geometry we cannot define *the* line segment connecting two points in a consistent way. Two points divide the line passing through them in *two* simple arcs, and there is no consistent way to distinguish one from the other. It is therefore impossible to tell whether a point r lies between two given points p, q.

- *Directions are ambiguous.* By the same token, we cannot define *the* direction from point p to point q. In particular, each point at infinity lies simultaneously in two opposite directions, as seen from a finite point. This property often makes it hard to use points at infinity as "sentinels" in geometric algorithms and data structures.

- *There are no convex figures.* The notion of convex set has no meaning in projective geometry. The problem is not just that the standard definition of convex set ("one that contains every segment joining two of its points") becomes meaningless, but in fact that there is no consistent way to distinguish between convex and non-convex sets.

Of course, we can avoid all these problems by letting our definitions of segment, direction, and so on depend on a special line Ω. However, we would then have to exclude certain "degenerate" cases, such as segments with endpoints on Ω. The concepts thus defined will not be preserved by arbitrary projective maps and will have uninteresting duals. In fact, this "solution" means giving up on projective geometry, and retreating to the Euclidean world.

4. Oriented projective geometry

Oriented projective geometry retains most advantages of the classical theory, but avoids the problems listed in the previous section. Its primitive objects are points and *oriented flats*: oriented lines, oriented planes, and so on. In particular, every straight line has an intrinsic orientation, which determines a "forward" direction along the line at every one of its points.

Every line of the unoriented projective plane is thus replaced by two coincident but oppositely oriented (hence distinct) lines. In order to maintain the exact duality between points and lines, each point must also be replaced by two "oppositely oriented" copies. Algebraically, this means treating $[w, x, y]$, $[-w, -x, -y]$ as distinct points, and $\langle W, X, Y \rangle$, $\langle -W, -X, -Y \rangle$ as distinct lines. The resulting set of points is topologically a double covering of the projective plane. Accordingly, I will use *two-sided* as a synonym of *oriented projective*. The two-sided plane in fact has the topology of a sphere, with straight lines corresponding to oriented great circles. Therefore, two-sided geometry is simply an oriented version of spherical geometry.

The double covering makes it possible to postulate an intrinsic *circular orientation* for the whole plane, which defines the "positive sense of turning" at every point, in a consistent way. We can then talk about the orientation of other objects in absolute terms: we can say that a triangle is "positively oriented," without having to specify a reference triangle. The global orientation of the plane also makes it possible to use the "forward" direction of a line to define its "left" and "right" sides.

In oriented projective geometry, a pair of points p, q generally determines not one but *two* distinct lines, with the same position but opposite orientations. We still can unambiguously speak of *the* line through p to q, if we pay attention to the order of those two points, distinguishing between the line joining p to q and the one joining q to p. Dually, two lines l and m on the plane have generally two points in common, so we must distinguish the point where l meets m from the point where m meets l. We will see that the two points can be distinguished by taking into account the orientations of l and m, and the global orientation of the plane.

The previously mentioned advantages of projective geometry are retained in the oriented version. In particular, we are still able to define an exact duality between points and lines that preserves not only the incidence properties of all objects, but also their relative orientations. In addition, the oriented version allows us to define the concept of convexity in a truly projective way. Unlike the Cartesian definition, the new one is unaffected by arbitrary projective maps and duality: we can finally say that the problem of intersecting n half-planes is *exactly* dual to finding the convex hull of n points, and not *approximately* so. Indeed, the ability to support both convexity and duality is perhaps the greatest advantage of the new framework.

Oriented projective geometry extends quite nicely to spaces of dimension greater than two. Indeed, it gives us effective and reliable tools for reasoning about the orientation of objects in spaces of arbitrary dimension, where our geometric intuition is weak and unreliable. Oriented projective geometry can be viewed as the marriage of projective geometry with an *algebra of orientations*.

5. Related work

The idea of distinguishing homogeneous tuples that differ by a negative factor is not entirely new to computer graphics programmers: as a matter of fact, they routinely make this distinction when computing a perspective view of a solid object. One step in this process is applying to the whole three-space a projective transformation which keeps the screen plane fixed and moves the observer to infinity. (Algebraically, this operation consists of multiplying the homogeneous coordinates of every point by a 4×4 matrix.) This transformation turns all rays out of the observer's eye into parallel lines, so that the perspective view can be obtained by a simple parallel projection of the transformed scene. However, it also has the unwanted effect of "folding" those parts of the image that originally were behind the observer (and hence invisible to him) over the visible part of the image [16]. See figure 6.

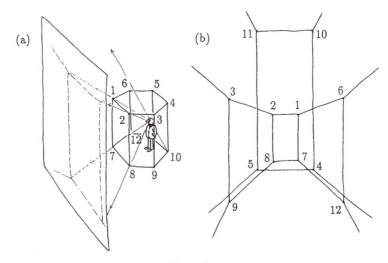

Figure 6.

It turns out that if the homogeneous coordinates of the original points are sign-normalized to have positive weight, then the invisible points (and only those) will have negative weight after the transformation. Although this fact is well-known and widely used by graphics programmers, it cannot be explained within the standard theory of homogeneous coordinates, which allows the transformation procedure to arbitrarily reverse the sign of all coordinates of the result. The negative-weight clipping rule above is therefore presented as a programming trick, based on implicit assumptions about the inner workings of the transformation procedure.

Graphics programmers also distinguish between homogeneous tuples of opposite sign when they test the sign of $Ww + Xx + Yy$ in order to determine whether the point $[w, x, y]$ is on the left or right side of the line $\langle W, X, Y \rangle$. This formula implicitly assumes that the line $\langle W, X, Y \rangle$ is distinct from the line $\langle -W, -X, -Y \rangle$, and also that $[w, x, y]$ is distinct from $[-w, -x, -y]$. To use this formula, the programmer must ensure that the coordinates of the point $[w, x, y]$ are "sign-normalized" in some consistent way (for example, so that $w > 0$), and that the coefficients $\langle W, X, Y \rangle$ have the desired sign. Besides the practical nuisance of requiring explicit tests and sign reversals, these normalization rules contradict the definition of homogeneous coordinates, destroy the exact duality between points and lines, and are mathematically inconsistent in many other ways, especially in the treatment of points at infinity [19].

On the theoretical front, Hermann Grassmann seems to have been the first to consider a geometric calculus based on two dual products (what we call join and meet), about a hundred years ago. His ideas were explored and reformulated by several other mathematicians since then, notably Clifford, Schröder, Whitehead, Cartan, and Peano. For a recent exposition of the ideas involved, see for example the paper by Berman [3] or the book by Hestenes and Sobczyk [13]. For some reason, the geometric calculus developed by those authors was relegated to relative obscurity, and its usefulness for practical computations has been largely ignored so far. Part of the reason may be the highly abstract language, excessive generality, and heavy mathematical formalism used in most expositions, which make the fundamental ideas seem much more complicated than they really are.

The notation used in this monograph is quite similar to the one used in a recent paper by Barnabei, Brini, and Rota [2], although it was developed independently from their work. The notion of an oriented flat as defined in the next chapters is closely related to what they call an *extensor*, or decomposable antisymmetric tensor. More precisely, the flats of oriented projective geometry are the equivalence classes we obtain by considering two extensors equivalent iff they differ by a positive scalar factor. Compared to their paper, this monograph gives more emphasis to the geometric (as opposed to algebraic) aspects of the calculus, and in particular to its suitability as the common language of computational geometry.

Chapter 2
Oriented projective spaces

As explained in the introduction, the concept of *oriented projective space* (or *two-sided space*) is best defined by means of concrete models. In this chapter I will describe a *canonical two-sided space* $\mathbf{T}_\nu$ for each dimension ν. This object will consist of an oriented manifold Υ_ν (whose elements are called *points*) and a collection $\mathcal{F}_\nu$ of oriented submanifolds of Υ_ν (the *flats*). I will then define a generic oriented projective space as any pair (U, F) isomorphic to $(\Upsilon_\nu, \mathcal{F}_\nu)$ for some ν.

Actually, I will construct three equivalent versions of $\mathbf{T}_\nu$, analogous to the straight, spherical, and analytic models of $\mathbf{P}_2$.

1. Models of two-sided space

1.1. The spherical model

The *spherical model* of $\mathbf{T}_\nu$ consists of the unit sphere $\mathbf{S}_\nu$ of $\mathbf{R}^{\nu+1}$, that is, the set of all points $(x_0, \ldots x_\nu)$ of $\mathbf{R}^{\nu+1}$ such that $\sum x_i^2 = 1$. Note that diametrically opposite points are *not* identified. For example, $\mathbf{T}_1$ (the *two-sided line*) is modeled by the unit circle of $\mathbf{R}^2$, and $\mathbf{T}_2$ (the *two-sided plane*) is modeled by the unit sphere of $\mathbf{R}^3$. See figure 1.

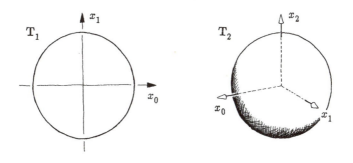

Figure 1. The spherical models of $\mathbf{T}_1$ and $\mathbf{T}_2$.

1.2. The straight model

The *straight model* of $\mathbf{T}_\nu$ consists of two copies of $\mathbf{R}^\nu$ (the *front* and *back ranges*), and one *point at infinity* $d\infty$ for every direction vector d in $\mathbf{R}^\nu$. (In this chapter, *direction* means a unit-length vector.)

For instance, the straight model of $\mathbf{T}_1$ consists of two copies of the real line $\mathbf{R}$, and two points at infinity $+\infty$ and $-\infty$. We can visualize this model as an infinite ruler with graduated scales on both sides. See figure 2.

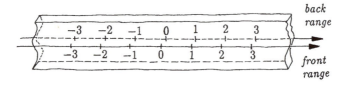

Figure 2. The straight model of $\mathbf{T}_1$.

Similarly, the straight model of $\mathbf{T}_2$ consists of two copies of $\mathbf{R}^2$, and an infinity point $d\infty$ for every direction d of $\mathbf{R}^2$. We can visualize the front and back ranges as two parallel planes in three space, or as the two sides of an infinite sheet of paper. Figure 3 is a sketch of this model, where the front and back ranges are represented by two copies of the open unit disk. The infinity point $d\infty$ is represented by point d on the boundary of the front disk, and point $-d$ on the boundary of the other.

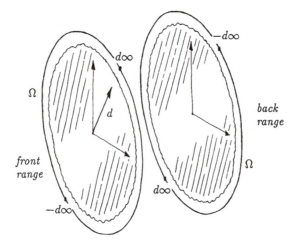

Figure 3. The straight model of $\mathbf{T}_2$.

This contracted version gives a more accurate picture of the topology of $\mathbf{T}_2$, in particular around the line at infinity. Note that the infinity points $d\infty$ and $(-d)\infty$ are *not* identified (unlike the conventions of standard projective geometry). Each infinity point is incident to both ranges, but in a rather peculiar way: by definition, $d\infty$ is a limit point of the front range in the direction d, and of the back range in the *opposite* direction $-d$.

The straight model suggests a convenient representation for figures on the two-sided plane: simply draw the front and back parts on the same sheet of paper, with coincident coordinate frames, using different graphical styles for each range. I will use solid dots, solid lines and cross-hatching for elements on the front range, and open dots, dashed lines, and dotted patterns for the back range. See figure 4.

Figure 4. Graphical conventions for the two-sided plane.

Figure 5 is a sketch of the straight model of three-dimensional oriented projective space $\mathbf{T}_3$, with each range contracted down to a copy of the unit open ball of $\mathbf{R}^3$. A point at infinity $d\infty$ is represented by point d on the boundary of the first ball, and point $-d$ on the boundary of the second.

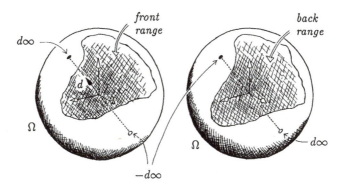

Figure 5. A sketch of the straight model of $\mathbf{T}_3$.

We can imagine the front half as being the set of "real" points, and the back half as a parallel universe of "phantom" points. Both ranges extend throughout the whole $\mathbf{R}^3$, but the only way to go from one to the other is through the points at infinity.

Recall that the *unoriented* projective space $\mathbf{P}_\nu$ can be constructed from $\mathbf{R}^\nu$, by adding to it one point at infinity for every pair of opposite directions $\{+d, -d\}$. The straight model of $\mathbf{T}_\nu$ is clearly a double covering of this construction.

1.3. The analytic model

The *analytic model* of $\mathbf{T}_\nu$ consists of the non-zero vectors of $\mathbf{R}^{\nu+1}$, where two vectors are considered to be the same point if one is a *positive* multiple of the other. I will denote by $[u]$ or $[u_0, .. u_\nu]$ the point represented by the vector $u = (u_0, .. u_\nu)$ and its positive multiples; any of those vectors is called the (*signed*) *homogeneous coordinates* of that point. Obviously, $[u_0, .. u_\nu] = [v_0, .. v_\nu]$ if and only if $u_i = \alpha v_i$ for all i and some *positive* real α. Note that $[u_0, .. u_\nu]$ and $[-u_0, .. -u_\nu]$ are distinct points of $\mathbf{T}_\nu$.

2. Central projection

The three models are related by *central projection* from the origin of $\mathbf{R}^{\nu+1}$. A point $[w, x, y, \ldots, z]$ of the analytic model corresponds to the points

$$\frac{(w, x, y, \ldots, z)}{\sqrt{w^2 + x^2 + y^2 + \cdots + z^2}}$$

of the spherical model and $(x/w, y/w \ldots, z/w)$ of the straight model. By definition, the latter is on the front range if $w > 0$, on the back range if $w < 0$, and at infinity in the direction $(x, y, \ldots, z)$ if $w = 0$.

2.1. Central projection of the two-sided line

In the case of $\mathbf{T}_1$, for example, central projection identifies the homogeneous pair $[w, x]$ with the point x/w of the front or back range of $\mathbf{T}_1$, depending on whether $w > 0$ or $w < 0$. The points $(0, 1)$ and $(0, -1)$ are mapped to $+\infty$ and $-\infty$, respectively.

Geometrically, this process can be described as follows. First, we draw the *front* range of $\mathbf{T}_1$ on the plane $\mathbf{R}^2$, as a vertical axis with its origin at the point $(1, 0)$; and we project the points of the *left* half of the circle onto this axis, by straight rays emanating from the origin $(0, 0)$. See figure 6(a). Next, we let that same vertical axis represent the *back* range of $\mathbf{T}_1$, and we project the points of the *right* half of the unit circle onto this axis, across the center, as shown in figure 6(b).

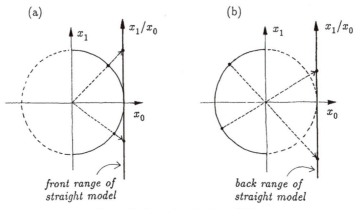

Figure 6. Central projection of $\mathbf{T}_1$.

Algebraically, central projection allows us to view a point $[w, x]$ of $\mathbf{T}_1$ as the fraction x/w, provided we distinguish it from the fraction $(-x)/(-w)$. The two fractions lie on different ranges, but have the same *numerical value*, that is, they have the same position within their range. As we will see in chapter 15, we can operate with these "two-sided fractions" in pretty much the same way we operate with normal ones.

2.2. Central projection of the two-sided plane

The two-dimensional case is entirely analogous. Imagine that the front range of $\mathbf{T}_2$ is embedded in $\mathbf{R}^3$, with the origin at $(1, 0, 0)$ and coordinate axes parallel to $(0, 1, 0)$ and $(0, 0, 1)$. See figure 7(a).

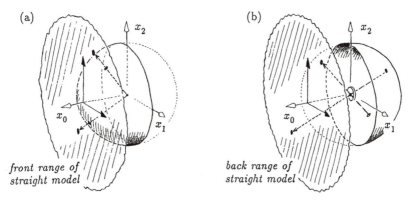

Figure 7. Central projection of $\mathbf{T}_2$.

Now imagine a light source at the center of the unit sphere $\mathbf{S}_2$. Central projection

takes every point of the "front" hemisphere of S_2 (i.e., with $w > 0$) to the shadow
it casts on the front face. For the other half of the mapping, we must place the
back range of T_2 in exactly the *same* position and orientation. Then points on the
back hemisphere of S_2 are projected onto it across the point $(0, 0, 0)$, as if through
a camera lens. See figure 7(b). Here are some examples:

$$[+3, +2, +6] \mapsto \left(+\tfrac{2}{3}, +\tfrac{6}{3}\right) \qquad \text{front}$$

$$[-3, -2, -6] \mapsto \left(+\tfrac{2}{3}, +\tfrac{6}{3}\right) \qquad \text{back}$$

$$[+3, -2, -6] \mapsto \left(-\tfrac{2}{3}, -\tfrac{6}{3}\right) \qquad \text{front}$$

$$[-3, +2, +6] \mapsto \left(-\tfrac{2}{3}, -\tfrac{6}{3}\right) \qquad \text{back}$$

$$[\ \ 0, +3, +4] \mapsto \left(+3, +4\right)\infty \qquad \text{infinity}$$

$$[\ \ 0, -3, -4] \mapsto \left(-3, -4\right)\infty \qquad \text{infinity}$$

Observe how the front hemisphere is merely stretched and flattened out by this
projection, whereas the back hemisphere suffers an additional 180° rotation.

2.3. Final comments

I will adopt central projection as the standard correspondence between the
three models, and generally think of them as the same mathematical object. In
definitions and theorems I will use whichever model is more convenient, and let
central projection implicitly carry the same concepts to the other two.

The analytic model is the most convenient to use in actual computations and
data structures. The other two are useful mainly as visual aids in the interpretation
of problems and the derivation of algorithms. The straight model is of course the link
between Euclidean and oriented projective geometry, since the front range of T_ν is a
faithful model of Euclidean space. The spherical model of T_1 and T_2 makes it easier
to visualize their topological and geometric properties (particularly at the infinity
points). Unfortunately, the spherical model of T_ν isn't nearly as useful for $\nu \geq 3$,
since the geometry of S_ν is then hard to visualize.

Chapter 3
Flats

The geometric structure of $\mathbf{T}_\nu$ is largely determined by its *flats*. These are sets of points roughly equivalent to the lines, planes, and higher-dimensional subspaces of classical geometry. One major difference is that every flat of $\mathbf{T}_\nu$ has an intrinsic *orientation*: a 1-dimensional flat is like a *directed* line, a 2-dimensional flat is like a plane with a built-in notion of "positive turn," and so on.

1. Definition

Definition 1. In the spherical model of $\mathbf{T}_\nu$, a *flat set* is a great sphere of $\mathbf{S}_\nu$, that is, the intersection of $\mathbf{S}_\nu$ and some linear subspace of $\mathbf{R}^{\nu+1}$. A *flat* is an oriented flat set.

A precise definition of "oriented" will be given in the next chapter; for now, it suffices to say that a great sphere can be oriented in exactly two ways. So, for every flat a there is an *opposite* flat $\neg a$, consisting of the same set of points (the same great sphere) taken with opposite orientation. Needless to say, I will always regard a and $\neg a$ as distinct flats. It is convenient to define $\sigma \circ f$, for any flat f and any $\sigma \in \{\pm 1\}$, as being f if $\sigma = +1$, and $\neg f$ if $\sigma = -1$.

As a special case of definition 1, I will postulate the existence of two flats of dimension -1, the *positive vacuum* Λ and its opposite, the *negative vacuum* $\neg\Lambda$. They should be regarded as oriented versions of the empty set.

Flats of dimension 1, 2, and 3 are called *lines*, *planes*, and *three-spaces*. There are only two flats with dimension ν, namely the universe Υ_ν and its oppositely oriented version $\neg\Upsilon_\nu$. Flats with dimension $\nu - 1$ are called *hyperplanes*.

2. Points

A flat set of dimension zero consists of two antipodal points of $\mathbf{S}_\nu$. As we will see in chapter 4, orienting such a set means designating one of the two points as the "positive" one. Therefore, the zero-dimensional oriented flats can be identified with the points of $\mathbf{T}_\nu$.

19

If p is a point, its opposite $\neg p$ is also called its *antipode*; it is the point diametrically opposite to p in the spherical model. See figure 1(a).

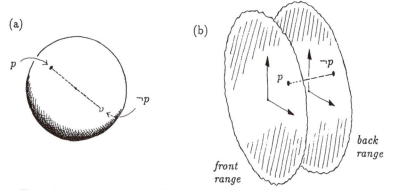

Figure 1. Antipodal points of $\mathbf{T}_2$ in the spherical and straight model.

In the straight model, the points p and $\neg p$ are either two points at infinity, in diametrically opposite directions, or two ordinary points with the same Cartesian coordinates, one on each range of $\mathbf{T}_\nu$. See figure 1(b). In the analytic model, the antipode of point $[w, x, y, .. z]$ is point $[-w, -x, -y, ..-z]$.

3. Lines

In the spherical model, a line of $\mathbf{T}_\nu$ is an oriented great circle of the unit sphere $\mathbf{S}_\nu$. The orientation can be visualized as an arrow that tells which direction along the circle is *positive* ("forward"). The opposite $\neg l$ of a line l is the same great circle with the arrow going the other way.

The one-dimensional space $\mathbf{T}_1$ has exactly two lines, namely the circle $\mathbf{S}_1$ taken in its two possible orientations. By definition, the universe Υ_1 of $\mathbf{T}_1$ is oriented counter-clockwise, that is, from $(1, 0)$ to $(0, 1)$ by the shortest route. See figure 2.

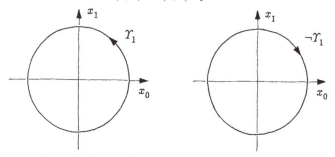

Figure 2. The two lines of $\mathbf{T}_1$ in the spherical model.

To understand what these two lines look like in the straight model, imagine a point p moving counterclockwise on $\mathbf{S}_1$, and consider what happens to its image under central projection. See figure 3.

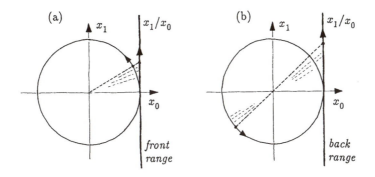

Figure 3.

While p is on the right half of $\mathbf{S}_1$, its image scans all points on the front range, in increasing order. As p goes through the point $(0, 1)$ at the top of the circle, its image becomes the infinity point $+\infty$, and then suddenly jumps to the back range, at large *negative* values. It then traverses the entire back range, *again in increasing order*. Finally, when p goes through the point $(0, -1)$, its image becomes the other infinity point $-\infty$, and then returns to the front range, again at the *negative* end.

Therefore, a cyclic ordering of the points of $\mathbf{S}_1$ corresponds in the straight model to either increasing or decreasing order of points on the front range, and the *same* ordering on the back range. See figure 4.

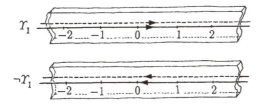

Figure 4. The two lines of $\mathbf{T}_1$ in the straight model.

3.1. Lines on the plane

Let's now consider lines in higher-dimensional spaces. In the spherical model of $\mathbf{T}_2$, a line is an oriented great circle of the units sphere of $\mathbf{R}^3$. See figure 5.

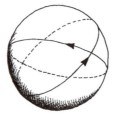

Figure 5. Lines in the spherical model of $\mathbf{T}_2$.

Central projection of such a great circle in the straight model may give either a a *proper* or an *improper* line. A proper line consists of two copies of the same directed Euclidean line, one on the front range and one on the back range, plus the two points at infinity incident to them. See figure 6. The two improper lines of $\mathbf{T}_2$ are the set of all of points at infinity, oriented in the two possible ways. The one oriented counterclockwise (as seen from the front range) is by definition the *horizon* Ω_2 of $\mathbf{T}_2$.

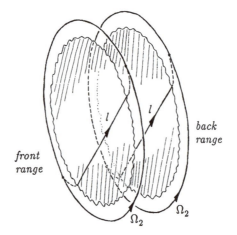

Figure 6. A proper line l and the horizon Ω_2 in the straight model of $\mathbf{T}_2$.

Note that the front and back parts of a proper line are directed the *same* way: the line always "moves" in the same direction d, in the front and in the back range.

3.2. Lines in three-space

In the straight model of $\mathbf{T}_3$ (and of higher-dimensional spaces as well), the situation is quite similar. An improper line is an oriented "great circle" of points at infinity, i.e. a circularly ordered set of points of the form $d\infty$, for all directions d lying on some plane of $\mathbf{R}^3$. (Unlike $\mathbf{T}_2$, $\mathbf{T}_3$ has infinitely many improper lines.)

A proper line consists of two copies of the same oriented Euclidean line, one on each range, and the two points at infinity incident to it. See figure 7.

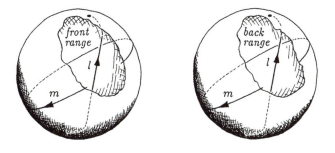

Figure 7. A proper line (l) and an improper line (m) in the straight model of $\mathbf{T}_3$.

Note how proper and improper lines of $\mathbf{T}_3$ look very much like the straight and spherical models of $\mathbf{T}_1$, respectively. This is true in general: in the straight model of $\mathbf{T}_\nu$, a κ-dimensional flat either looks like the spherical model of $\mathbf{T}_\kappa$, expanded to infinite radius, or looks like the straight model of $\mathbf{T}_\kappa$, embedded in the straight model of $\mathbf{T}_\nu$.

4. Planes

In the spherical model, a plane (two-dimensional flat) is an oriented great 2-sphere of $\mathbf{S}_\nu$. The orientation can be visualized as a small "circular arrow" painted on the sphere. By sliding this arrow along the sphere's surface, we can tell whether a turn at any point is *positive* (agreeing with the arrow) or *negative*. The opposite $\neg\pi$ of a plane π is the same great 2-sphere with the arrow turning the other way.

The two-dimensional space $\mathbf{T}_2$ contains exactly two planes, the universe Υ_2 and its opposite $\neg\Upsilon_2$. By definition, Υ_2 has the circular arrow at $(1,0,0)$ turning from direction $(0,1,0)$ to direction $(0,0,1)$ by the shortest angle. If the axes of $\mathbf{R}^3$ are arranged in space in the usual way, then Υ_2 is the sphere $\mathbf{S}_2$ oriented counterclockwise, as seen from the outside. See figure 8.

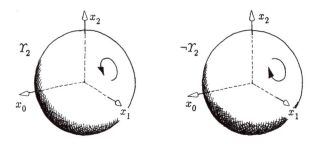

Figure 8. The two planes of $\mathbf{T}_2$.

What does the orientation of Υ_2 look like in the straight model? If we consider how central projection affects the direction of turns at various points, we see that positive turns of Υ_2 become counterclockwise turns on the front range, and *clockwise* turns on the back range. See figure 9.

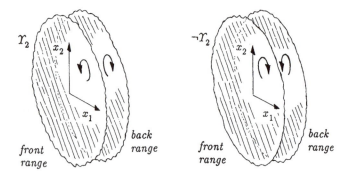

Figure 9. The two planes of $\mathbf{T}_2$ in the straight model.

4.1. Planes in three-space

In the straight model of $\mathbf{T}_3$, a plane may be either *proper* or *improper*. A proper plane consists of two oppositely oriented copies of the same Euclidean plane, one on each range of $\mathbf{T}_\nu$, and all points at infinity in directions parallel to that plane. See figure 10.

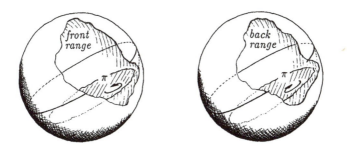

Figure 10. A proper plane π of $\mathbf{T}_3$.

In $\mathbf{T}_3$ there are exactly two oppositely oriented improper planes, consists of all the points at infinity (which are topologically equivalent to a two-dimensional sphere). By definition, the *celestial sphere* Ω_3 of $\mathbf{T}_3$ is the improper plane that is oriented *clockwise* as seen from the front range, and counter-clockwise as seen from the back range. See figure 11.

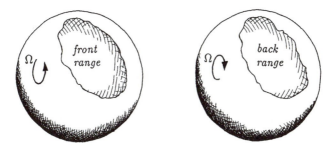

Figure 11. The celestial sphere Ω_3 of $\mathbf{T}_3$.

In spaces of dimensions higher than three the situation is similar, except that there are infinitely many improper planes.

5. Three-spaces

A three-space (three-dimensional flat) is an oriented great 3-sphere of $\mathbf{S}_\nu$. We can visualize the orientation of a three-space as a "corkscrew" arrow, or as a combination of a circular arrow and a straight arrow perpendicular to it. We can also depict the orientation as a tiny hand, with the stretched thumb replacing the straight arrow and the curled fingers replacing the curved arrow. See figure 12.

Figure 12. Three-dimensional orientations.

Two such devices represent the same orientation if we can transform one into the other by a continuous rigid motion without leaving the three-space. Therefore, the difference between the two orientations of a three-space correspond to the difference between a left hand and a right hand, or between two oppositely-threaded screws.

Recall that in the straight model a three-space consists of two copies of $\mathbf{R}^3$, plus a sphere of points at infinity. If we consider what happens to a smoothly moving corkscrew arrow as crosses the sphere at infinity, we will find that the orientation of the three-space induces the *same* orientation in the front and back ranges.

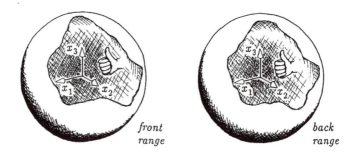

Figure 13. The universe Υ_3 of $\mathbf{T}_3$.

In particular, the space $\mathbf{T}_3$ contains only two three-dimensional flats, the standard universe Υ_3 and its opposite $\neg\Upsilon_3$. See figure 13. By definition, the orientation of Υ_3 is given by a screw arrow that turns from $(1,0,0)$ towards $(0,1,0)$ while moving from $(0, 0, 0)$ towards $(0, 0, 1)$. If the coordinate axes of $\mathbf{R}^3$ are depicted according to the usual mathematical conventions, this orientation is given by thumb and fingers of the *right* hand.

It is worth emphasizing that in the straight model, a line or a three-space has both ranges oriented the same way, whereas a plane has the two ranges oriented in opposite ways. The general rule will be given in the following chapter.

6. Ranks

In projective geometry, it is often convenient to classify spaces and flats by their *rank*, defined to be their dimension plus one. Thus, for example, the vacuum has rank 0, points have rank 1, lines have rank 2, planes have rank 3, and so on. As we shall see, formulas that deal with dimensions often become much simpler if expressed in terms of ranks.

To keep formulas short and reduce the possibility of confusion, I will adopt the following convention: the Greek letters $\kappa, \mu, \nu, \rho, \sigma, \tau$ will usually denote dimensions, and the corresponding italic letters k, m, n, r, s, t will usually denote the corresponding ranks. The identities $k = \kappa + 1$, $m = \mu + 1$, and so on will be assumed throughout. With this convention we can say, for example, that $\mathbf{S}_\nu$ is the unit sphere of $\mathbf{R}^n$, and a κ-dimensional great sub-sphere of $\mathbf{S}_\nu$ is the intersection of $\mathbf{S}_\nu$ and a k-dimensional linear subspace of $\mathbf{R}^n$.

I will denote the set of all flats of rank k in $\mathbf{T}_\nu$ by $\mathcal{F}^k_\nu$, or simply $\mathcal{F}^k$ when the dimension ν is clear from the context.

7. Incidence and independence

By definition, two flats of $\mathbf{T}_\nu$ are *incident* to each other if the corresponding flat sets (in the spherical model) have non-empty intersection; otherwise, they are *independent* of each other. In particular, a point p is incident to a line l (and vice versa) if and only if the two antipodal points of $\mathbf{S}_\nu$ corresponding to p belong to the great circle corresponding to l.

Note that every point p is incident to both itself and to its antipode $\neg p$, and is independent of any other point. Note also that Λ and $\neg \Lambda$ are independent from every flat, including themselves.

Chapter 4
Simplices and orientation

In the chapter 3 a κ-dimensional flat of $\mathbf{T}_\nu$ was defined as an oriented κ-dimensional great sphere of $\mathbf{S}_\nu$. It is now time to define precisely what "oriented" means in this context.

It is possible to define the concept of orientation for an arbitrary manifold in purely topological terms, but this approach requires some heavy mathematical machinery and would take us far from the focus of this book. Fortunately, for the particular manifolds we are interested here in (great spheres of $\mathbf{S}_\nu$) we can get by with a much simpler definition, based on elementary linear algebra.

1. Simplices

A *simplex* is an ordered tuple of points of $\mathbf{T}_\nu$, called the *vertices* of the simplex. In the spherical model, a simplex is an ordered tuple of unit vectors of $\mathbf{R}^n$. The simplex is *proper* if those vectors are linearly independent; otherwise it is *improper*, or *degenerate*.

Let us consider some examples. A two-vertex simplex is an ordered pair of points p, q. See figure 1. This simplex is degenerate if and only if the two points are equal or antipodal.

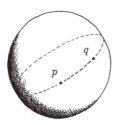

Figure 1. A two-vertex simplex of $\mathbf{T}_2$.

A three-vertex simplex is degenerate if and only if its vertices lie on the same great circle of $\mathbf{S}_\nu$, that is, are coplanar vectors of $\mathbf{R}^n$; that is to say, the three points lie on the same line of $\mathbf{T}_\nu$. See figure 2.

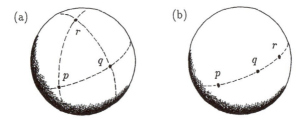

Figure 2. Three-vertex simplices of $\mathbf{T}_2$: (a) proper, (b) degenerate.

For any simplex s of $\mathbf{T}_\nu$ there is a unique flat set of minimum dimension that contains s, the so-called *span* of s. In the spherical model, the span of s is the great sphere of $\mathbf{S}_\nu$ determined by the linear subspace of $\mathbf{R}^n$ generated by the vertices of s (viewed as vectors). The *dimension* of s is by definition that of its span. Obviously, a simplex with k vertices is proper if and only if its dimension is $\kappa = k-1$. Thus, for example, the span of a proper simplex with two vertices is the unique great circle of $\mathbf{S}_\nu$ that contains those two points.

2. Simplex equivalence

I will say that two proper simplices of $\mathbf{T}_\nu$ are *equivalent* if we can continuously transform one into the other in such a way that all intermediate stages span the same unoriented flat. For example, two proper simplices with three vertices are equivalent if they lie on the same unoriented plane of $\mathbf{T}_\nu$, and we can continuously move one onto the other without leaving that plane and without ever making the three points collinear.

2.1. Equivalence of bases

This notion of simplex equivalence is closely related to that of *basis equivalence* in a real vector space V. Two bases of V are said to be equivalent if it is possible to continuously deform one into the other, in such a way that every intermediate stage is a basis of V.

Observe that a proper simplex that spans a great sphere C is a basis for the subspace V of $\mathbf{R}^n$ that defines C. Conversely, from any basis of V we can get a proper simplex that spans C by rescaling the basis vectors to unit length. Since the vectors of a basis must have nonzero length, this map from bases to simplices is continuous. We conclude that two proper simplices are equivalent if and only if they are equivalent bases of the same subspace of $\mathbf{R}^n$.

The condition for two bases to be equivalent is a well-known result of elementary linear algebra. In order to state this condition concisely, it is convenient to view an ordered sequence s of k vectors (in particular, a k-vertex simplex) as a matrix

$$s = \begin{pmatrix} s^0 \\ s^1 \\ \vdots \\ s^\kappa \end{pmatrix}$$

whose rows are the given vectors. (To save space, I will often write such a matrix also as $(s^0; s^1; \ldots s^\kappa)$, using semicolons instead of commas to denote vertical stacking.) With this convention, we can say that two sequences of k vectors u, v span the same vector space V if and only if there is a $k \times k$ matrix A such that $Au = v$. Also, if u and v are bases of V, the matrix A is unique and has a non-zero determinant; and, furthermore, the bases are equivalent if and only if the determinant is positive.

2.2. Orientations as simplex classes

By the same token, the simplices spanning a given great sphere (flat set) C of $\mathbf{S}_\nu$ are also divided into two equivalence classes. I will identify these two classes with the two *orientations* of C. By naming one of these classes the set of *positive* bases, we get an *oriented great sphere* of $\mathbf{S}_\nu$, i.e. an (oriented) flat of $\mathbf{T}_\nu$. Therefore, a proper κ-dimensional simplex $s = (s^0; \ldots s^\kappa)$ determines a unique flat of $\mathbf{T}_\nu$, namely the smallest flat set containing s, oriented so that s is a positive simplex. I will denote this flat by $[s]$, by $[s^0; \ldots s^\kappa]$, or by the matrix

$$\begin{bmatrix} s_0^0 & \cdots & \cdots & s_\nu^0 \\ \vdots & & & \vdots \\ s_0^\kappa & \cdots & \cdots & s_\nu^\kappa \end{bmatrix}$$

where $s^i = [s_0^i, \ldots s_\nu^i]$.

2.3. Orientation of a point

To make these notions clear, let's have a look at the low-dimensional cases. According to the definition, a zero-dimensional flat set is an unordered pair of antipodal points of the sphere. An *oriented* zero-dimensional flat is one such pair, with one of the points singled out as the "positive" simplex. Obviously, zero-dimensional flats can be identified with the points of $\mathbf{T}_\nu$.

2.4. Orientation of two points on a line

A one-dimensional flat set of $\mathbf{T}_\nu$ is a great circle C of $\mathbf{S}_\nu$. An orientation for this set is a class of equivalent non-degenerate one-dimensional simplices.

Consider for example two proper simplices $(p;q)$ and $(r;s)$ of C. The two simplices are equivalent if and only if we can continuously move p to r and q to s, without leaving C, and without going through a degenerate state. At every instant during this motion the simplex $(p;q)$ determines a unique circular ordering of the points of C, namely the one that goes from p to q by the shortest route; this circular ordering can be depicted as a longitudinal arrow on C. See figure 3.

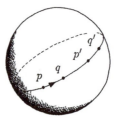

Figure 3.

This ordering remains unchanged while p and q move around, as long as $p \neq q$ and $p \neq \neg q$. It follows that $(p;q)$ and $(r;s)$ are equivalent only if they define the same circular ordering on C. The two orientations of C correspond to its two possible circular orderings. Note that the proper simplex $(p;q)$ is equivalent to $(\neg p; \neg q)$ but not to $(q;p)$ or $(\neg p;q)$.

2.5. Orientation of a triangle

Let's now consider simplices with three vertices. A proper simplex $(p;q;r)$ defines a unique two-dimensional great sphere C, and a unique spherical triangle on C. See figure 4.

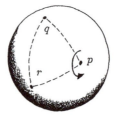

Figure 4. Orientation of simplex $(p;q;r)$.

The sides of the triangle are the shortest arcs of great circle connecting each pair of vertices. Note that for the simplex to be proper no two vertices may be coincident or antipodal.

We can visualize the orientation of the simplex as a small circular arrow surrounding the point p, turning from the direction of pq to that of pr by the shortest angle. Note that the angle between the arcs pq and pr at p cannot be zero or $180°$, since in that case the three vertices would be coplanar vectors of $\mathbf{R}^n$. If the three points move continuously on C in such a way that they always form a proper simplex, then the circular arrow is well defined at all times, and simply slides over C following the point p. Two simplices s and t will be equivalent only if the circular arrow determined by s at s^0 can be transported over C so as to coincide with that defined by t at t^0. In particular, observe that the proper simplex $(p; q; r)$ is equivalent to the cyclically permuted copies $(q; r; p)$ and $(r; p; q)$, but not to $(r; q; p)$, $(q; p; r)$, or $(\neg p; \neg q; \neg r)$.

2.6. Orientation of a tetrahedron

A proper three-dimensional simplex $s = (o; p; q; r)$ defines six "edges", the shortest arcs of great circle connecting each pair of vertices. The orientation of the simplex can be visualized as a small corkscrew arrow, near the point o, that turns from the direction of edge op to that of oq by the shortest angle, while at the same time advancing towards r along the edge or. Alternatively, we can imagine a small hand at the point o, with the thumb pointing towards r and the other fingers curled in the sense from op to oq. See figure 5.

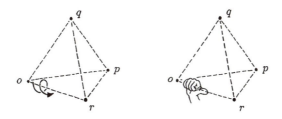

Figure 5. Orientation of simplex $(o; p; q; r)$.

This corkscrew arrow stays well-defined during any continuous deformation of the simplex, as long as the four points do not become coplanar. If the motion is confined to a three-dimensional great sphere C, a simplex with left-threaded screw cannot be deformed into one with a right-threaded screw.

In particular, the simplex $s = (o; p; q; r)$ is equivalent to $(q; r; o; p)$ and $(\neg o; \neg p; \neg q; \neg r)$, but not to $(p; q; r; o)$ or $(\neg o; p; q; r)$.

2.7. Orientation of the universe

The canonical basis of $\mathbf{R}^n$ defines the *standard simplex* $\mathbf{e} = (\mathbf{e}^0; \ldots \mathbf{e}^\nu)$ of $\mathbf{T}_\nu$. The point $\mathbf{e}^0 = [1, 0, \ldots, 0]$ is the *front origin*, also denoted by O, and $\mathbf{e}^1$ through $\mathbf{e}^\nu$ are the *cardinal directions*. By definition, the universe Υ_ν of $\mathbf{T}_\nu$ is oriented so that this standard simplex is positive. For example, the universe of $\mathbf{T}_2$ has the orientation of the simplex with vertices $[1, 0, 0]$, $[0, 1, 0]$, and $[0, 0, 1]$. See figure 6. In the straight model, the standard simplex consists of the "vertices" of the first quadrant of the front range: the origin, the infinity point on the x-axis, and the infinity point on the y-axis, in that order.

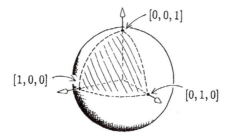

Figure 6. The standard simplex of $\mathbf{T}_2$.

3. Point location relative to a simplex

3.1. Segments

In section 2.4 we mentioned the shortest great circle arc connecting two distinct and non-antipodal points p, q of $\mathbf{T}_\nu$. It is quite natural to define *the segment pq* as being the set of points on this arc. In other words, x is on the (open) segment pq if and only if the simplices $(p; x)$ and $(x; q)$ are proper and equivalent to $(p; q)$. This set is empty if $p = q$ or $p = \neg q$. See figure 7.

3.2. The interior of a simplex

We can generalize the notion of segments to higher dimensions as follows. If $s = (s^0; \ldots s^\kappa)$ is a proper κ-dimensional simplex of $\mathbf{T}_\nu$, then we define the *interior of simplex s* as the set of all points x which produce a proper simplex equivalent to s when substituted for any of its vertices. That is, x is in the interior of s if and only if the simplex $(s^0; \ldots s^{i-1}; x; s^{i+1}; \ldots s^\kappa)$ is equivalent to s for all i.

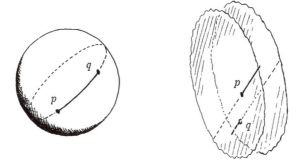

Figure 7. The segment pq.

In particular, the interior of a proper three-vertex simplex $(p; q; r)$ is the set of all points x such that the turns from xp to xq, from xp to xr, and from xr to xp are in the same direction as the turn from pq to pr. I will call this set the (*open*) *triangle pqr*. See figure 8.

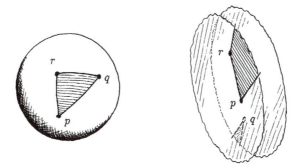

Figure 8. The triangle pqr.

3.3. Locating a point in a simplex

More generally, if x is a point on the flat spanned by a simplex $s = (s^0; \ldots s^\kappa)$, we can classify the position of x with respect to s by substituting x for each vertex of s in turn, and comparing the orientation of the result with that of s. The outcomes of those tests can be represented by a string $\sigma_0 \sigma_1 \ldots \sigma_\nu$ of signs: $\sigma_i = +1$ (or simply '+') if replacing the ith vertex by x produces an equivalent simplex, $\sigma_i = '-'$ if it produces one with opposite orientation, and $\sigma_i = 0$ if it produces a degenerate simplex. This sequence is the *signature* of x relative to the simplex.

Any proper simplex s therefore partitions $\mathbf{T}_\nu$ into regions, each of them consisting of the points with the same signature relative to s. Of the 3^k possible sign sequences, all but one correspond to a non-empty region; the only signature that does not apply to any point is the null one, $000\cdots0$. In particular, the signature $+++\cdots+$ corresponds to the interior of the simplex. In general, a signature $\sigma_0\cdots\sigma_\nu$ with no '0's denotes the interior of the simplex $(\sigma_0\!\circ\!s^0,\ \sigma_1\!\circ\!s^1,\ ..\ \sigma_\nu\!\circ\!s^\nu)$. A point x whose signature σ has only $m < k$ non-zero signs lies on the subflat defined by the corresponding m vertices of s. In that case, if $t = [s^{i_0}; s^{i_1}; ..\, s^{i_\mu}]$ is the sub-simplex of s consisting of those vertices, in the same order as they appear in s, then the signature of x relative to t will be $\sigma_{i_0}\sigma_{i_1}\cdots\sigma_{i_\mu}$. Thus, for example, the signature $+0+00\cdots0$ means x is in the flat spanned by s^0 and s^2, and its signature with respect to $[s^0; s^2]$ is $++$; in other words, x is in the open segment $s^0 s^2$.

For example, consider the proper simplex (p,q,r) of figure 9. The great circles determined by each pair of vertices cut the plane $\mathbf{T}_2$ into $3^3 - 1 = 26$ regions: eight open triangles, twelve open segments, and six isolated points. The signature $+++$ is the interior of the triangle pqr; $---$ denotes its antipodal image, the triangle $\neg p\,\neg q\,\neg r$. Signatures $++0$, $+-0$, $--0$, and $-+0$ stand for the open segments pq, $p\,\neg q$, $\neg p\,\neg q$, $\neg p\,q$. The signature $+00$ is produced only by the point p itself, and -00 only by its antipode $\neg p$. And so on.

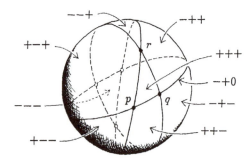

Figure 9. Some signatures relative to the simplex (p, q, r).

3.4. Computing signatures

Computing the signature by the definition given above would require computing the sign of k simplices, that is, computing k determinants of order $k \times k$. In practice, a much better method is to decompose the homogeneous coordinates of x into a linear combination of the coordinates of the s^i; the signs of the coefficients of this linear combination are the desired signature. That is, if $x = [x_0, \ldots x_\nu]$ and $s^i = [s_0^i, \ldots s_\nu^i]$, we must solve the linear system of equations

$$(\alpha_0, \cdots \alpha_\kappa) \begin{pmatrix} s_0^0 & s_1^0 & \cdots & s_\nu^0 \\ \vdots & & \vdots & \\ s_0^\kappa & s_1^\kappa & \cdots & s_\nu^\kappa \end{pmatrix} = (x_0, x_1, \cdots x_\nu) \tag{1}$$

for the unknowns $\alpha_0, \ldots \alpha_\kappa$, and take $\sigma_i = \operatorname{sign}(\alpha_i)$. The correctness of this algorithm is a trivial exercise in linear algebra. Incidentally, observe that by formula (1) the signature $00 \cdots 0$ can be realized only by the null homogeneous tuple $[0, 0, \ldots 0]$, which is not a point of $\mathbf{T}_\nu$.

In particular, if we want to test whether a point x of $\mathbf{T}_1$ is in the segment determined by two given points p, q, we must solve

$$(\alpha_0, \alpha_1) \begin{pmatrix} p_0 & p_1 \\ q_0 & q_1 \end{pmatrix} = (x_0, x_1) \tag{2}$$

which means computing the three determinants

$$\delta = \begin{vmatrix} p_0 & p_1 \\ q_0 & q_1 \end{vmatrix} = p_0 q_1 - p_1 q_0$$

$$\beta_0 = \begin{vmatrix} x_0 & x_1 \\ q_0 & q_1 \end{vmatrix} = x_0 q_1 - x_1 q_0$$

$$\beta_1 = \begin{vmatrix} p_0 & p_1 \\ x_0 & x_1 \end{vmatrix} = p_0 x_1 - p_1 x_0$$

Then the signature of x relative to $(p; q)$ will be $(\operatorname{sign}(\delta)\operatorname{sign}(\beta_0), \operatorname{sign}(\delta)\operatorname{sign}(\beta_1))$.

4. The vector space model

When looking at simplices of $\mathbf{S}_\nu$ as bases of $\mathbf{R}^n$, the fact that their vectors have unit length is an irrelevant complication. We can get rid of this assumption by working with a fourth model of the $\mathbf{T}_\nu$.

Let's define an *oriented vector space* as a vector space with one of its two classes of equivalent bases singled out as the *positive* class. In the *vector space model* of $\mathbf{T}_\nu$ a κ-dimensional flat is represented by a $(\kappa + 1)$-dimensional oriented subspace of $\mathbf{R}^n$. In this model, a point (0-dimensional flat) is represented by a one-dimensional oriented subspace of $\mathbf{R}^n$. The orientation of that subspace (the set of its positive bases) consists of all positive multiples of a single vector u. We can depict that subspace as a directed line through the origin of $\mathbf{R}^n$, pointing in the same direction as u. See figure 10.

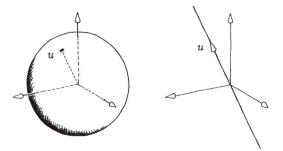

Figure 10. A point of $\mathbf{T}_2$ in the spherical and vector space models.

A line of $\mathbf{T}_2$ is represented by a two-dimensional linear subspace of $\mathbf{R}^3$. This subspace can be visualized as a plane passing through the origin of $\mathbf{R}^3$, with a circular arrow on it. See figure 11.

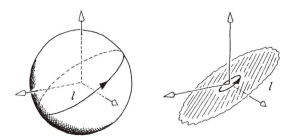

Figure 11. A line of $\mathbf{T}_2$ in the spherical and vector space models.

The arrow shows the direction of the shortest turn from the first to the second vector of any positive basis of that subspace. That arrow agrees with the orientation of the great circle of $\mathbf{S}_2$ representing the same line in the spherical model. In the vector space model, the universe Υ_ν of $\mathbf{T}_\nu$ is represented by the space $\mathbf{R}^n$ itself, with the canonical basis $\mathbf{e}^0, \ldots \mathbf{e}^\nu$ taken as positive.

Chapter 5
The join operation

In classical projective geometry, the join of two flats is defined as the smallest flat containing them. Thus, for instance, the join of two points is the line passing through them, and the join of a point and a line is the plane that contains both. The join operation in oriented projective geometry is quite similar, except that its arguments and its result are *oriented* flats. Therefore, its definition must say how the orientation of the result is determined from that of the arguments.

1. The join of two points

Let's consider first the join of two points in the spherical model of $\mathbf{T}_\nu$. Two points p, q generally determine a unique great circle of $\mathbf{S}_\nu$, and divide it into two unequal arcs. The *segment pq* is, by definition, the shorter of these two arcs. If we orient that great circle from p to q along the segment pq, we get the *join of p to q*, denoted by $p \vee q$. See figure 1.

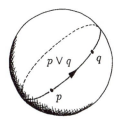

Figure 1. The join of two points.

This definition is meaningful if and only if the two points are independent, that is, $p \neq q$ and $p \neq \neg q$. Observe that $q \vee p$ is oriented in the direction opposite to that of $p \vee q$. That is, the join of two points is *anticommutative*: for all independent p, q,

$$q \vee p = \neg(p \vee q)$$

Observe also that the shortest arcs from p to q and from p to $\neg q$ leave p in opposite directions. Similarly, the arcs from p to q and from $\neg p$ to q arrive at q from opposite directions. Therefore, for all independent pairs p, q,

$$p \vee (\neg q) = \neg(p \vee q) = (\neg p) \vee q$$

1.1. Join in the straight model

What do the segment pq and the join $p \vee q$ look like in the straight model? If both p and q are points at infinity, the join is either $\neg\Omega$ or Ω, depending on whether the shortest turn from the direction of p to that of q (as seen from the front range) is clockwise or counterclockwise.

If at least one of the points is finite, their join is a proper line of $\mathbf{T}_\nu$, that is, two copies of the same directed Euclidean straight line (one on each range), each passing through p or $\neg p$ and through q or $\neg q$. Each of the two great circle arcs connecting p and q corresponds to some subset of those two lines and their infinity points. Note that central projection does not preserve arc length, but we still can recognize the "shorter" of the two arcs as the one which does not contain any antipodal pairs. In particular, if p and q both lie on the same range, the segment pq coincides with its Euclidean definition, and the line $p \vee q$ is oriented the obvious way. See figures 2(a) and 2(b).

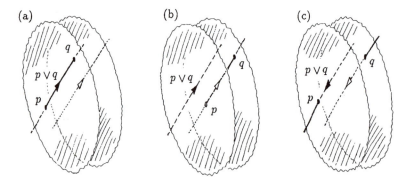

Figure 2. Join in the straight model.

If the two points are on opposite ranges, as in figure 2(c), the line $p \vee q$ is directed *away* from $\neg q$ at p, and *towards* $\neg p$ at q. Observe that the path from p to q in this direction, consisting of the ray leaving p and the one ending at q, is indeed the shortest one. The alternate path, consisting of the two complementary rays, includes all points in the Euclidean segment from p to $\neg q$ and their antipodes on the segment from $\neg p$ to q.

If p is finite but q is not, then the line $p \vee q$ is directed from p to q along the ray connecting the two. Note that if $q = d\infty$ and p is on the back range, then the direction of the line will be $-d$. See figures 3(a) and 3(b).

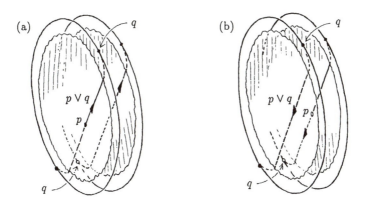

Figure 3. Join with infinity points, in the straight model.

2. The join of a point and a line

Let l be a line, and p be a point not on l. The *join of p to l*, by definition, is the plane of $\mathbf{T}_\nu$ that contains both and is oriented so that l turns around p in the positive sense. See figure 4.

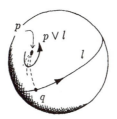

Figure 4. The join of p to l.

More precisely, let a point q move forwards on l, and observe the sense in which the line $p \vee q$ turns at l; by definition, that is the orientation of $p \vee l$.

Clearly, the orientation of a point-line join depends on the orientation of both operands:

$$p \vee (\neg l) = (\neg p) \vee l = \neg (p \vee l) \tag{1}$$

The symmetric operation $l \vee p$ can be defined as either the same as $p \vee l$ or as its opposite. It is desirable to define it in such a way as to make $\vee$ associative; that is,

$$p \vee (q \vee r) = (p \vee q) \vee r \tag{2}$$

for all $p, q, r \in \mathcal{P}$ for which those operations are defined. It turns out that both sides of equation (2) are undefined on exactly the same cases, namely when p, q, and r lie on a common line. Equations (1) and (2) imply

$$
\begin{aligned}
p \vee (q \vee r) &= (p \vee q) \vee r \\
&= \neg((q \vee p) \vee r) \\
&= \neg(q \vee (p \vee r)) \\
&= q \vee (r \vee p) \\
&= (q \vee r) \vee p
\end{aligned}
\tag{3}
$$

Since every line l can be expressed as the join of some two points q and r, we conclude that

$$p \vee l = l \vee p \tag{4}$$

for every point p and every line l. Therefore, in order to make $\vee$ associative, we must make it commutative in the point-line case. At first sight this may seem a poor choice, considering that join is anti-commutative in the point-point case. Actually, the two definitions are quite consistent with each other. Derivation (3) above essentially says that to go from $p \vee l$ to $l \vee p$ we must reverse *two* point-point joins, and therefore the orientation of the result is not affected.

3. The join of two arbitrary flats

Observe that the join of points p and q can be defined as the line containing both, oriented so that the pair p, q is a positive simplex. Moreover, if p is a point, l is a line, and $(q; r)$ is a positive simplex of l, then the join of p to l is the plane containing both, oriented so that $(p; q; r)$ is a positive simplex. I will define the join of general flats by generalizing this observation. That is,

Definition 1. The join of two flats determined by simplices u, v is the flat defined by their concatenation. That is,

$$[u^0; \ldots u^\kappa] \vee [s^0; \ldots s^\mu] = [u^0; \ldots u^\kappa; s^0; \ldots s^\mu]$$

With a little linear algebra we can easily check that the concatenation of two proper simplices is a proper simplex if and only if the corresponding flats have no point in common. If they do, their join is undefined. It is easy to see also that the result of the join is the same no matter which simplices we choose to represent the two flats. The join of flats with rank zero must be defined separately. By definition,

$$\Lambda \vee a \;=\; a \;=\; a \vee \Lambda$$
$$(\neg\Lambda) \vee a \;=\; \neg a \;=\; a \vee (\neg\Lambda)$$

for all a. In other words, Λ is the (left and right) identity of join. Note that the vacua Λ and $\neg\Lambda$ are disjoint from every flat, even from themselves; they behave like oriented empty sets — hence their names.

4. Properties of join

Note that every flat of rank $k \geq 1$ is the join of the k vertices of any of its positive simplices. Obviously, whenever $a \vee b$ is defined we have

$$\mathrm{rank}(a \vee b) \;=\; \mathrm{rank}(a) + \mathrm{rank}(b) \tag{5}$$

Also,

$$(\neg a) \vee b \;=\; a \vee (\neg b) \;=\; \neg(a \vee b) \tag{6}$$

for all disjoint flats a, b.

4.1. Associativity

The associativity of $\vee$ follows directly from the definition: we have

$$a \vee (b \vee c) \;=\; (a \vee b) \vee c \tag{7}$$

for any flats a, b, c, when either side is defined.

4.2. Commutativity

Recall that the join of two points depends on the order of the operands, whereas that of a point and a line does not. The general rule follows readily from definition 1. Observe that transposing the order of two vectors in a basis reverses its orientation. Observe also that in going from $[a^0; \ldots a^\kappa; b^0; \ldots b^\mu]$ to $[b^0; \ldots b^\mu; a^0; \ldots a^\kappa]$ we have to transpose $(\kappa + 1)(\mu + 1)$ adjacent pairs of vectors. We conclude that, for any two flats a, b,

$$a \vee b \;=\; \neg^{\mathrm{rank}(a)\,\mathrm{rank}(b)}(b \vee a) \tag{8}$$

That is, reversing the order of the operands in a join reverses the orientation of the result as many times as the product of the ranks of those operands. Therefore, $b \vee a$ is $a \vee b$ if one of the operands has even rank, and $\neg(a \vee b)$ if both have odd rank. Observe that equations (6), (7), and (8) are valid even for flats of rank zero.

5. Null objects

Recall that $a \vee b$ is undefined if a and b have a common point. It may be tempting to plug this hole, and extend the definition so that $a \vee b$ is always one of the two smallest flats containing a and b. This extension is common in classical projective geometry, but unfortunately it cannot be made to work in the oriented framework. If a and b are not disjoint, it is impossible to define the orientation of $a \vee b$ in a consistent way.

We can understand the difficulty as follows. If we apply the commutativity law (8) to the expression $p \vee p$, we get

$$p \vee p \;=\; \neg(p \vee p)$$

which cannot be true for any point, or indeed for any flat. This shows we cannot consistently define $p \vee p$. In general, if two flats a and b have a common point p, we can always write them as $a = u \vee p$ and $b = p \vee v$ for some (possibly vacuous) flats u, v. Then by associativity we must have

$$a \vee b \;=\; u \vee (p \vee p) \vee v$$

We conclude that $a \vee b$ cannot be consistently defined when a and b are not disjoint.

This problem is not as serious as it may seem. Even in unoriented geometry the extended join operation cannot be made continuous, since the dimension of the result may change abruptly with infinitesimal changes in the operands. In most geometry algorithms, those degenerate cases require special handling anyway, so the proposed extension would not make programs much simpler. The extended join may actually be a nuisance in strongly typed languages such as Pascal and Modula-2, where one usually wants to use different compile-time types for geometric objects of different dimensions.

Nevertheless, from the programmer's viewpoint partially defined operations are bothersome. It is generally preferable to turn them into total functions, by adding a new "undefined" element to their range, and letting this element be the result of the operation whenever it was not defined originally. To this end, I will postulate a dummy *null object* $\mathbf{0}^k$ for every integer k, and let $a \vee b$ be $\mathbf{0}^{\text{rank}(a)+\text{rank}(b)}$ whenever a and b are not disjoint.

As we shall see, this extension is quite natural from the computational point of view, and can be implemented at zero or negative cost: the same code that produces $a \vee b$ in the normal case will produce the null object $\mathbf{0}^k$ if a and b are not disjoint. In fact, a practical test for whether a and b intersect is to compute $a \vee b$ by the standard algorithm, and check whether the result is $\mathbf{0}^k$.

It is convenient to define also $\neg \mathbf{0}^k = \mathbf{0}^k$, and $\mathbf{0}^k \vee a = a \vee \mathbf{0}^k = \mathbf{0}^k \vee \mathbf{0}^m = \mathbf{0}^{k+m}$, for all flats a of rank m. With these rules, all properties of join listed so far are always true, even when the operands are not disjoint and/or are null objects. However, note that $\mathbf{0}^k$ is quite unlike ordinary flats in many respects; for example, it fails to satisfy $a \neq \neg a$. For that reason, I will neither call it a flat nor include it in $\mathcal{F}$.

Usually, the rank of null objects is irrelevant or known from the context, so I will write simply $\mathbf{0}$ instead of $\mathbf{0}^k$. However, in computer implementations (especially in strongly-typed languages) it is usually simpler to to use different representations for the null object of each rank.

6. Complementary flats

An important result, that is easily proved by reference to the vector space model, is

Theorem 1. *For any subflat x of a flat f there is a flat y such that $x \vee y = f$.*

PROOF: Let X and F be the oriented vector spaces representing x and f in the vector space model. Since x is a subflat of f, the space X is a subspace of F. From linear algebra we know that there is a basis $s = (s^0; \ldots s^\kappa)$ for F whose first m elements ($m = $ rank of x) are a basis for X.

Let u be the flat determined by the first m elements of s, and v be the one determined by the last $k - m$. Because of the way s was selected, u is either x or $\neg x$, and $u \vee v$ is either f or $\neg f$. It follows that either $x \vee v = f$ or $x \vee (\neg v) = f$. QED.

A flat y with this property is said to be a *right complement of x in f*. (Symmetrically, the flat x is said to be a *left complement* of x in f.) Note that the (right and left) complement of f itself in f is Λ, and that of $\neg f$ is $\neg \Lambda$; and vice-versa. Except for these special cases, the right and left complements are not unique.

Chapter 6
The meet operation

Besides join, the other fundamental operation of classical geometry is computing the the point of intersection, or *meet*, of two given lines on the plane; or, more generally, computing the flat that is the intersection of two given flats.

In classical (unoriented) geometry, flats are considered to be sets of points, and therefore their intersection doesn't have to be defined explicitly: it is plain set intersection. In the oriented version this simple definition is not enough, since the meet operation must also choose an orientation for the resulting set. In this chapter we will see how to select this orientation in a consitent fashion. The meet operation thus defined has properties quite similar to those of join. In fact, in Chapter 10 we will see that meet and join are dual operations, in a very precise sense.

1. The meeting point of two lines

Two lines of $\mathbf{T}_2$ generally intersect on a pair of antipodal points. See figure 1. To choose an orientation for the intersection means to pick one member of the pair as *the* meeting point of the two lines.

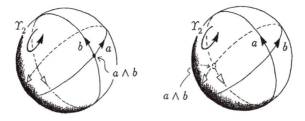

Figure 1. The meet of two lines.

47

Note that the shortest turn from the direction of a to that of b is positive at one of the two common points, and negative at the other. By definition, the first one is the *point where a meets b*, denoted by $a \wedge b$. More precisely, if u_a and u_b are vectors tangent to a and b (and agreeing with their directions) at $p = a \wedge b$, then p, u_a, and u_b (in that order) are a positive basis of $\mathbf{R}^3$. Note that at the antipodal point all three vectors are exactly reversed, and therefore form a basis of opposite handedness. The meet $a \wedge b$ is not defined when $a = b$ or $a = \neg b$.

Like the join of two points, the meet of two lines is anticommutative, and depends on the orientation of its operands. For any two lines a, b of $\mathbf{T}_2$, we have

$$b \wedge a \;=\; \neg(a \wedge b)$$

$$(\neg a) \wedge b \;=\; a \wedge (\neg b) = \neg(a \wedge b)$$

This behavior is unavoidable if the meet operation is to be continuous and defined for any two unrelated lines. To see why, consider two lines a and b (on the spherical model of $\mathbf{T}_2$) that intersect at a point p in such a way that b is 90° counterclockwise from a. See figure 2(a).

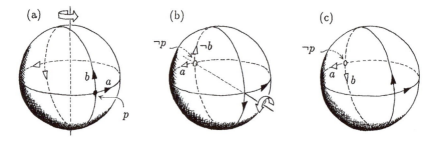

Figure 2.

Imagine that the sphere rotates by 180° around an axis perpendicular to the plane of a. This continuous motion keeps a fixed, but takes b to $\neg b$ and exchanges p with $\neg p$. See figure 2(b). Therefore, no matter which of the two points we define to be $a \wedge b$, we must have $a \wedge (\neg b) = \neg(a \wedge b)$.

Now start from the situation in figure 2(b), and rotate the sphere by 90° around the line of the two intersection points, so as to take a into b and $\neg b$ into a. See figure 2(c). By continuity, we conclude that $b \wedge a = a \wedge (\neg b)$, which we have just shown to be $\neg(a \wedge b)$.

1.1. The relativity of meet

Note that the meet of two lines as defined above depends strongly on the orientation of the whole plane $\mathbf{T}_2$. This dependency is not matter of choice, but rather an essential property of oriented intersections. It turns out that it is not possible to consistently select one of the intersection points without a reference orientation for the whole plane.

To see why, consider two intersecting finite lines a, b of $\mathbf{T}_3$ (in the straight model). Let l be the bisecting line of the angle ab. Now rotate a and b by 180° around the axis l. This continuous motion exchanges a with b, while keeping the intersection points fixed and avoiding degeneracies ($a = b$ or $a = \neg b$). By continuity, we should then have $b \wedge a = a \wedge b$. However, this conclusion contradicts the previous proof that $b \wedge a = \neg(a \wedge b)$. Therefore, if a and b are intersecting lines in three-space, any definition of $a \wedge b$ must be ambiguous, or must be discontinuous for some pairs a, b with $a \neq b$ and $a \neq \neg b$.

Therefore, we cannot define the oriented meet of two coplanar lines in $\mathbf{T}_3$ or in a higher-dimensional space, since (as we saw in the previous chapter) there is no consistent way to pick an orientation for the plane containing them. In general, the meet of two flats cannot be defined in an absolute sense, but only relative to some oriented flat of suitable dimension that contains them.

2. The general meet operation

The meeting point of two lines in $\mathbf{T}_2$ can also be defined as folows: for any three points p, q, r of $\mathbf{T}_2$,

$$p \vee q \vee r = \Upsilon_2 \;\Leftrightarrow\; (p \vee q) \wedge (q \vee r) = q \tag{1}$$

See figure 3.

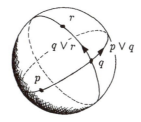

Figure 3. The meet of two lines in $\mathbf{T}_2$.

We can define the meet of two flats of arbitrary dimension by a straightforward extension of formula (1):

Definition 1. If u is a flat of minimum rank enclosing flats a and b, then the *meet of a and b in u* is the flat y such that

$$x \vee y \vee z = u$$
$$x \vee y = a \tag{2}$$
$$y \vee z = b$$

for some $x, z \subseteq u$. I will denote the meet y by $a \wedge_u b$.

In other words, for any three flats x, y, z such that $x \vee y \vee z = u \neq \mathbf{0}$, we have by definition

$$(x \vee y) \wedge_u (y \vee z) = y$$

Before we go on, we must prove that definition 1 is consistent: that is, we must show that such a flat y always exists and is unique. The next lemma shows this is indeed the case:

Theorem 1. *If u is a flat of minimum rank containing flats a and b, then there are some flats x, z, and a unique flat y satisfying equations* (2).

PROOF: The intersection of a and b, viewed as sets of points, is some unoriented flat contained in u. Let y be any oriented version of that flat. Since $y \subset a$, it has a left complement in a: that is, there is flat x such that $x \vee y = a$. Similarly, there is a flat z such that $y \vee z = b$.

Since z is contained in b and disjoint from $y = a \cap b$, we conclude z is disjoint from a, and therefore the flat $v = x \vee y \vee z = a \vee z$ is well-defined. Now x, y, and z are contained in u, so v too is contained in u. Also, v contains $x \vee y = a$ and $y \vee z = b$. Since u is a flat of minimum rank containing a and b, v must be either u or $\neg u$. Then either the flats x, y, z or the flats $\neg x, \neg y, \neg z$, respectively, will satisfy (2).

To show the uniqueness of y, let x, y, and z be defined as above, with orientations reversed as needed to make equations (2) hold. Suppose that equations (2) are satisfied also by flats x', y', z'. Since $a = x' \vee y'$ and $b = y' \vee z'$, we must have $y' \subseteq a \cap b$, and $x' \subseteq a \setminus b$. Since $a \cap b$ is the set of points y, we conclude

$$y' \subseteq y \tag{3}$$

We also conclude $x' \cap y = \emptyset$, which means $x' \vee y$ is defined; since this flat is contained in a, we must have $\mathrm{rank}(x') + \mathrm{rank}(y) \leq \mathrm{rank}(a)$. From this observation, and from $\mathrm{rank}(x') + \mathrm{rank}(y') = \mathrm{rank}(x' \vee y') = \mathrm{rank}(a)$ we conclude $\mathrm{rank}(y') \geq \mathrm{rank}(y)$. Together with (3) this implies $y' = y$ or $y' = \neg y$.

Now suppose $y' = \neg y$. Then from $x \vee y = a = x' \vee y'$ and $y \vee z = b = y' \vee z'$ we get

$$x \vee y \vee z = x' \vee y' \vee z$$
$$= x' \vee (\neg y) \vee z$$
$$= \neg(x' \vee y \vee z)$$
$$= \neg(x' \vee y' \vee z')$$

contradicting the assumption that $x \vee y \vee z = x' \vee y' \vee z'$.
QED.

Incidentally, this proof shows also that the meet of two flats is simply the intersection of their point sets, oriented in a specific way.

I will omit the subscript in $\wedge_u$ when the reference flat u (the *universe* of the operation) is implied by the context. In particular, I will use the phrases *the meet operation of* $\mathbf{T}_\nu$ or *the ν-dimensional meet* to signify the meet operation relative to the standard universe Υ_ν of $\mathbf{T}_\nu$.

2.1. Null objects

Note that definition 1 specifies $a \wedge_u b$ only if u is the smallest flat enclosing both a and b; which is to say, if and only if $\mathrm{rank}(a) + \mathrm{rank}(b) - \mathrm{rank}(a \cap b) = \mathrm{rank}(u)$. When that is not the case, $a \wedge_u b$ is undefined. As in the case of join, it is convenient to extend meet into a total function, by letting $a \wedge b$ be the null object $\mathbf{0}^{\mathrm{rank}(a)+\mathrm{rank}(b)-\mathrm{rank}(u)}$, whenever the result is not specified by definition 1. It is also convenient to postulate $\mathbf{0}^k \wedge a = a \wedge \mathbf{0}^k = \mathbf{0}^{k+\mathrm{rank}(a)-\mathrm{rank}(u)}$, for all a.

As in the case of join, the standard (unoriented) algorithms for computing the intersection of two flats can be modified to produce a correctly orientated result, and to return $\mathbf{0}$ according to the above rules, without increasing their cost or complexity.

3. Meet in three dimensions

3.1. The meet of a line and a plane

To clarify the definition of meet even further, let's consider some examples in $\mathbf{T}_3$. For instance, a line l and a plane π generally have two antipodal points in common. According to the definition, $\pi \wedge l = x$ if and only if there are points p, q on π and r on l such that $(p; q; x)$ is a positive triangle of π, $(x; r)$ is a positive pair on l, and $(p; q; x; r)$ is a positive tetrahedron of $\mathbf{T}_3$. We conclude that $\pi \wedge l$ is the point where the circular arrow of π and the longitudinal arrow of l are like the fingers and thumb of the right hand. See figure 4.

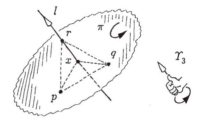

Figure 4. The meet of a line and a plane.

By an entirely analogous argument, we conclude that $l \wedge \pi$ is the same point as $\pi \wedge l$; that is, the meet of a line and a point is commutative. The result of $\pi \wedge l$ is undefined if l is contained in π.

3.2. The meet of two planes

The intersection of two planes π, σ in $\mathbf{T}_3$ is a pair of oppositely oriented lines. According to definition 1, we must find points p, q, r, s such that $(p; q; r)$ is a positive triangle of π, $(q; r; s)$ is a positive triangle of σ, and $(p; q; r; s)$ is a positive tetrahedron of $\mathbf{T}_3$. Then $\pi \wedge \sigma$ will be the line from q to r. See figure 5.

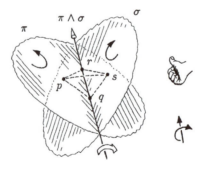

Figure 5. The meet of two planes in $\mathbf{T}_3$.

Informally, we must imagine π turning towards σ around their common line, by the smallest angle that makes the two planes coincide in position and orientation. Then we can figure out the direction of $\pi \wedge \sigma$ from this turning direction by the right-hand rule. Note that $\sigma \wedge \pi = \neg(\pi \wedge \sigma)$; that is, the meet of two planes in $\mathbf{T}_3$ is anticommutative.

4. Properties of meet

Note that the orientation of the result depends on that of the reference flat u, as well as on those of a and b. In fact, if we replace x, y, and z by their opposites in the equations (2), we conclude that

$$a \wedge_{(\neg u)} b = \neg(a \wedge_u b)$$

In what follows $\wedge$ denotes the meet operation relative to some fixed ν-dimensional flat Υ (which may or may not be the standard universe of $\mathbf{T}_\nu$).

4.1. Orientation reversal

The definition of meet implies that, for any flats or null objects a, b,

$$(\neg a) \wedge b = a \wedge (\neg b) = \neg(a \wedge b)$$

4.2. Meet with universe

The reference flat Υ acts as the unit element of meet: for all flats $a \subseteq \Upsilon$,

$$a \wedge \Upsilon = a = \Upsilon \wedge a$$
$$a \wedge (\neg \Upsilon) = \neg a = (\neg \Upsilon) \wedge a$$

4.3. Meet in different spaces

The relationship between the meet operation of different universes is illuminated by the next lemma, which follows immediately from the definition:

Lemma 2. *Let x, y, and u be flats such that $x \vee u \vee y \neq \mathbf{0}$. Then for any $a, b \subseteq u$ we have*

$$(x \vee a) \wedge_{x \vee u \vee y} (b \vee y) = a \wedge_u b$$

In particular, by taking $x = \Lambda$ or $y = \Lambda$ we conclude that, if $x \vee u \neq \mathbf{0}$, then

$$(x \vee a) \wedge_{x \vee u} b = a \wedge_u b \tag{4}$$

$$a \wedge_{u \vee x} (b \vee x) = a \wedge_u b \tag{5}$$

for all $a, b \subseteq u$. Among other things, this fact allows us to establish a connection between the meet operations in $\mathbf{T}_2$ and $\mathbf{T}_3$. Let's consider $\mathbf{T}_2$ embedded as a plane of $\mathbf{T}_3$ in the standard way: that is, let's identify the point $[w, x, y]$ of $\mathbf{T}_2$ with $[w, x, y, 0]$ of $\mathbf{T}_3$. With these conventions we can write $\mathbf{T}_3 = \mathbf{T}_2 \vee \mathbf{e}^3$, where $\mathbf{e}^3 = [0, 0, 0, 1]$ is the point at plus infinity on the front z-axis of $\mathbf{T}_3$. Then, by lemma 2, the meet of lines a and b (relative to $\mathbf{T}_2$) is the meet of the line a and the vertical plane $b \vee \mathbf{e}^3$ (relative to $\mathbf{T}_3$). See figure 6.

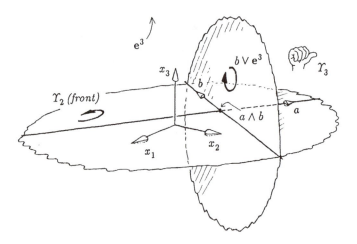

Figure 6. Relation between meet operation in $\mathbf{T}_2$ and $\mathbf{T}_3$.

4.4. Co-ranks

When working with the meet operation relative to a fixed universe Υ, it is convenient to classify the flats of Υ by their *complementary rank* or *co-rank*, defined by $\mathrm{corank}(a) = \mathrm{rank}(\Upsilon) - \mathrm{rank}(a) = \dim(\Upsilon) - \dim(a)$. In particular, the reference flat Υ itself has co-rank 0, and its hyperplanes have co-rank 1. The vacuum has co-rank equal to the rank of Υ. In general, the co-rank of a flat a is how many points must be joined to a to get the universe. From the definition of meet it follows that

$$\mathrm{corank}(a \wedge b) = \mathrm{corank}(a) + \mathrm{corank}(b)$$

and

$$\mathrm{rank}(a \wedge b) = \mathrm{rank}(a) - \mathrm{corank}(b) = \mathrm{rank}(b) - \mathrm{corank}(a)$$

That is, the meet operation lowers the rank of one operand by the co-rank of the other. Since hyperplanes of Υ have co-rank equal to one, we conclude that the co-rank of a flat a is the number of hyperplanes we have to meet in order to get a.

4.5. Commutativity

The general meet operation is either commutative or anti-commutative, depending on the ranks of the operands and the rank of the reference space. More precisely,

Theorem 3. *For all flats a and b,*

$$b \wedge a = \neg^{\mathrm{corank}(a)\,\mathrm{corank}(b)}(a \wedge b)$$

whenever a $\wedge$ b is defined.

PROOF: If either side is the null object, then the theorem is trivial. Otherwise, there must be flats x, y, z such that $x \vee y \vee z = \Upsilon$, $x \vee y = a$, and $y \vee z = b$, with $y = a \wedge b$. Let r, s, and t be the ranks of x, y and z, respectively. Then $z \vee y \vee x = \neg^{rs+rt+st}\Upsilon$. It follows that $(z \vee y) \wedge (y \vee x) = \neg^{rs+rt+st}y$. But $z \vee y = \neg^{st}b$, and $y \vee x = \neg^{rs}a$. Therefore, $b \wedge a = \neg^{rt}(a \wedge b)$. Since $r + s + t = \mathrm{rank}(\Upsilon)$ and $r + s = \mathrm{rank}(a)$, it follows that $r = \mathrm{corank}(b)$. Similarly $t = \mathrm{corank}(a)$, and the proof is complete. QED.

Theorem 3 says that $\wedge$ is commutative, unless both operands have odd co-rank. In three dimensions and less, the only "odd" cases are: two points in $\mathbf{T}_1$, two lines in $\mathbf{T}_2$, two planes or a point and a plane in $\mathbf{T}_3$.

4.6. Associativity

Like join, meet is associative. In order to prove this fact, we will need the following useful lemma, which provides an alternative to definition 1:

Lemma 4. *Let u be any flat. Then, for all $x, y, z \subseteq u$,*

$$\begin{aligned} x \vee y \vee z = u \; &\Leftrightarrow \; (x \vee y) \wedge_u (x \vee z) = x \\ &\Leftrightarrow \; (x \vee z) \wedge_u (y \vee z) = z \end{aligned} \qquad (6)$$

PROOF: Let $r = \mathrm{rank}(x)$, $s = \mathrm{rank}(y)$. By the commutativity properties of join and by definition 1, we have

$$\begin{aligned} x \vee y \vee z = u \; &\Leftrightarrow \; y \vee x \vee z = \neg^{rs}u \\ &\Leftrightarrow \; (y \vee x) \wedge (x \vee z) = \neg^{rs}x \\ &\Leftrightarrow \; (x \vee y) \wedge (x \vee z) = x \end{aligned}$$

The proof for $(x \vee z) \wedge_u (y \vee z) = z$ is entirely symmetric. QED.

It is important to notice that not every permutation of x, y, and z in the right-hand side of equation (6) will make the formulas true. For example, if $x \vee y \vee z = u$, it doesn't follow that $(y \vee z) \wedge_u (x \vee z) = z$.

Lemma 5. *For any flats a, b, c,*

$$(a \wedge b) \wedge c = \mathbf{0} \quad \Leftrightarrow \quad a \wedge (b \wedge c) = \mathbf{0} \tag{7}$$

PROOF: If either of a, b, or c is a null object, the theorem is obvious. Otherwise, if $a \wedge b = \mathbf{0}$, then a and b must be contained in some flat f with rank less than n; in that case, a and $b \wedge c$ are also contained in f, and therefore both sides of equation (7) are null. The case $b \wedge c = \mathbf{0}$ is entirely symmetrical. So, let's assume $a \wedge b$ and $b \wedge c$ are both defined; we must have

$$\mathrm{rank}(a) + \mathrm{rank}(b) - \mathrm{rank}(a \wedge b) = n \tag{8}$$
$$\mathrm{rank}(b) + \mathrm{rank}(c) - \mathrm{rank}(b \wedge c) = n \tag{9}$$

Subtracting (9) from (8) we get

$$\mathrm{rank}(a) + \mathrm{rank}(b \wedge c) = \mathrm{rank}(a \wedge b) + \mathrm{rank}(c)$$

Hence, the equality

$$\mathrm{rank}(a) + \mathrm{rank}(b \wedge c) - \mathrm{rank}(a \cap (b \wedge c)) = n$$

will be true if and only if

$$\mathrm{rank}(a \wedge b) + \mathrm{rank}(c) - \mathrm{rank}((a \wedge b) \cap c) = n$$

which means $a \wedge (b \wedge c)$ is defined if and only if $(a \wedge b) \wedge c$ is defined. QED.

Lemma 6. *For any flats a, b, c,*

$$(a \wedge b) \wedge c = \Lambda \quad \Leftrightarrow \quad a \wedge (b \wedge c) = \Lambda$$

PROOF: Suppose $(a \wedge b) \wedge c = \Lambda$. By lemma 5, $a \wedge (b \wedge c)$ must be defined. Since meet is an oriented intersection, and intersection is associative, $a \wedge (b \wedge c)$ must be either Λ or $\neg \Lambda$. Also, $a \wedge c$ must be defined; for, if there were some flat of less than full rank containig a and c, it would also contain $a \wedge b$ and c, and $(a \wedge b) \wedge c$ would be $\mathbf{0}$.

So, let's define

$$x = a \wedge b$$
$$y = a \wedge c$$
$$z = b \wedge c$$

I claim that $x \vee y = a$, $x \vee z = b$, $y \vee z = c$, and $x \vee y \vee z = \Upsilon$. First, since $x \cap y = a \cap b \cap c = \emptyset$, the flat $x \vee y$ is defined. Since x and y are flats of a, $x \vee y \subseteq a$ Also, since $\text{rank}((a \wedge b) \wedge c) = 0$, we deduce $\text{corank}(c) = \text{rank}(a \wedge b)$, and

$$\begin{aligned}
\text{rank}(x \vee y) &= \text{rank}((a \wedge b) \vee (a \wedge c)) \\
&= \text{rank}(a \wedge b) + \text{rank}(a \wedge c) \\
&= \text{corank}(c) + (\text{rank}(a) - \text{corank}(c)) \\
&= \text{rank}(a)
\end{aligned}$$

which means $x \vee y = \alpha \circ a$ for some $\alpha \in \{\pm 1\}$. In the same way, we can show that $x \vee z = \beta \circ b$ and $y \vee z = \gamma \circ c$ for some $\gamma \in \{\pm 1\}$.

The hypothesis $(a \wedge b) \wedge c = \Lambda$ implies $(a \wedge b) \vee c = \Upsilon$, and, therefore,

$$x \vee y \vee z = (a \wedge b) \vee (\gamma \circ c) = \gamma \circ (a \wedge b) \vee c = \gamma \circ \Upsilon \qquad (10)$$

By the definition of meet and by lemma 4 it follows from (10) that

$$\begin{aligned}
(x \vee y) \wedge (y \vee z) &= \gamma \circ y = \gamma \circ (a \wedge c) \\
(x \vee y) \wedge (x \vee z) &= \gamma \circ x = \gamma \circ (a \wedge b) \\
(x \vee z) \wedge (y \vee z) &= \gamma \circ z = \gamma \circ (b \wedge c)
\end{aligned} \qquad (11)$$

On the other hand,

$$\begin{aligned}
(x \vee y) \wedge (y \vee z) &= (\alpha \circ a) \wedge (\gamma \circ c) = \alpha \gamma \circ (a \wedge c) \\
(x \vee y) \wedge (x \vee z) &= (\alpha \circ a) \wedge (\beta \circ b) = \alpha \beta \circ (a \wedge b) \\
(x \vee z) \wedge (y \vee z) &= (\beta \circ b) \wedge (\gamma \circ c) = \beta \gamma \circ (b \wedge c)
\end{aligned} \qquad (12)$$

Comparing (11) and (12) we conclude $\alpha = \beta = \gamma = +1$. Therefore we have $x \vee y = a$ and $x \vee y \vee z = \Upsilon$, which means

$$a \wedge (b \wedge c) = (x \vee y) \wedge z = \Lambda$$

The converse follows from this equality and from the commutativity law (theorem 3).

QED.

We are now ready to prove the main result:

Theorem 7. *Meet is associative: for any three flats a, b, c,*

$$(a \wedge b) \wedge c = a \wedge (b \wedge c) \tag{13}$$

whenever one of the two expressions is defined.

PROOF: If one side of (13) is **0**, then by lemma 5 the other side is **0**, too. So, assume both sides are defined. Let $w = (a \wedge b) \wedge c$. Since w is a subflat of a, b, and c, there are flats t, u, v such that

$$a = w \vee t$$
$$b = w \vee u$$
$$c = w \vee v$$

Also, there is a flat s such that $w \vee s = \Upsilon$. Therefore, by equation (4) we have

$$(a \wedge b) \wedge c = \big((w \vee t) \wedge_{w \vee s} (w \vee u)\big) \wedge_{w \vee s} (w \vee v)$$
$$= \big(w \vee (t \wedge_s u)\big) \wedge_{w \vee s} (w \vee v)$$
$$= w \vee \big((t \wedge_s u) \wedge_s v\big)$$

Similarly,

$$a \wedge (b \wedge c) = w \vee (t \wedge_s (u \wedge_s v)) \tag{14}$$

From $w = (a \wedge b) \wedge c$ and equation (14) we conclude that $(t \wedge_s u) \wedge_s v = \Lambda$. By lemma 6, this fact implies $t \wedge_s (u \wedge_s v) = \Lambda$. Then equation (14) says

$$a \wedge (b \wedge c) = w \vee \Lambda = w = (a \wedge b) \wedge c$$

QED.

Chapter 7
Relative orientation

One advantage of two-sided geometry is that it allows us to talk about the two sides of a line in the plane, or of a plane in three-space. As shown below, these concepts can be elegantly expressed in terms of the join and meet operations.

1. The two sides of a line

A line l divides the spherical model of the two-sided plane in two halves. I call these the *left* and *right* (or *positive* and *negative*) sides of l, as they would be seen by an ant crawling along the line on the outside of the sphere. See figure 1.

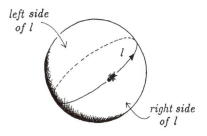

Figure 1. The two sides of a line of $\mathbf{T}_2$.

More precisely, let $(q; r)$ be any positive simplex of the line. I will say that a point p is on the *positive* (or *left*) *side* of l if the simplex $(p; q; r)$ (in that order) is a positive triangle of $\mathbf{T}_2$. Symmetrically, p is on the *negative* (*right*) side of l if $(p; q; r)$ is a negative triangle.

1.1. The sides of a line in the straight model

How are these concepts mapped to the straight model? If the line m is at infinity, then its left and right sides are the front and back ranges of $\mathbf{T}_2$ (when $m = \Omega$), or vice-versa (when $m = \neg\Omega$). If m is a finite line, the picture is a bit more involved. Let d be the direction of m, let L^+ and R^+ be the left and right half-planes determined by m on the front range, and let L^-, R^- be their antipodal images on the back range. Then the left side of m is the union of L^+, R^-, and the points at infinity $u\infty$ for u in the counterclockwise arc from $+d$ to $-d$. The right side of m consists of R^+, L^-, and the supplementary arc on the line at infinity. See figure 2. The reversal of left and right on the back range is to be expected, since the meaning of "counterclockwise" is reversed there, but the longitudinal orientation of the line l is not.

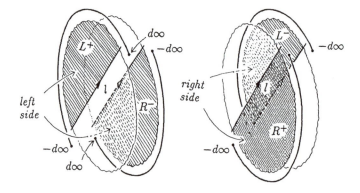

Figure 2. The two sides of a line, in the straight model of $\mathbf{T}_2$.

2. Relative position of arbitrary flats

Observe that we can express the left and right sides of a line in terms of the join and meet operations, as follows:

$$p \text{ is } \left\{ \begin{array}{c} \text{to the left of } l \\ \text{on } l \\ \text{to the right of } l \end{array} \right\} \text{ iff } p \vee l = \left\{ \begin{array}{c} \Upsilon_2 \\ 0^2 \\ \neg\Upsilon_2 \end{array} \right\} \text{ iff } p \wedge l = \left\{ \begin{array}{c} \Lambda \\ 0^0 \\ \neg\Lambda \end{array} \right\} \quad (1)$$

Notice how these formulas derive the "transversal" (left-right) orientation of l by combining the "longitudinal" orientation of l with the intrinsic "circular" orientation of the universe Υ_2.

The same idea can be used to distinguish the two sides (half-spaces) determined by a hyperplane of $\mathbf{T}_\nu$, for arbitrary ν. In fact, we can generalize (1) even further. Given two flats a and b of $\mathbf{T}_\nu$ whose ranks add to n, I define the *relative orientation of a and b* as the sign-valued function

$$a \diamond b = \begin{cases} +1 & \text{if } a \vee b = \ \Upsilon_\nu \\ 0 & \text{if } a \vee b = \ \mathbf{0}^n \\ -1 & \text{if } a \vee b = \neg\Upsilon_\nu \end{cases}$$

From the definitions of $\wedge$ and $\diamond$, it is easy to see that

$$a \diamond b = \begin{cases} +1 & \text{if } a \wedge b = \ \Lambda \\ 0 & \text{if } a \wedge b = \ \mathbf{0}^0 \\ -1 & \text{if } a \wedge b = \neg\Lambda \end{cases}$$

If we ignore the distinction between 1 and Λ, we can say that $a \diamond b$ is just a special case of $a \wedge b$, in which the two flats have complementary ranks. If $a \diamond b = +1$, I will say that *a is positively oriented with respect to b*, or that *the pair a, b is positively oriented*.

The operation $a \diamond b$ is defined if and only if the flats have complementary ranks (that is, $\text{rank}(a) + \text{rank}(b) = \text{rank}(\Upsilon)$), and is $\mathbf{0}$ if and only if a and b have a common point. Observe also that $a \diamond b$ depends on the orientation of Υ, as well as on those of a and b.

As in any join, reversing the order of the arguments reverses the sign of $a \diamond b$ if and only if both have odd rank:

$$b \diamond a = (-1)^{\text{rank}(a)\,\text{rank}(b)}(a \diamond b)$$

Since in this case $\text{rank}(a) + \text{rank}(b) = n$, we conclude that the order of the arguments only matters if the space has even rank (odd dimension) and one of the operands has odd rank (even dimension). In three dimensions or less, the only cases where the order of a and b matters are two points on a line, and a point versus a plane in three-space.

2.1. Signed predicates

The relative orientation symbol $\diamond$ is one of many sign-valued functions that are common in oriented projective geometry. The corresponding functions in unoriented geometry have only two outcomes, and are usually implemented as predicates, that is, procedures returning a boolean result. For example, the undirected analogue of $\diamond$ would be the predicate that tests whether a given point lies on a given line.

In oriented projective geometry, it is generally better to implement a function like $\diamond$ as a procedure returning an integer value in $\{-1, 0, +1\}$. This procedure can be used both in two-way if statements, as in if $\text{Rel}(p, l) = 0$ then ..., and in three-branched case statements. Experience seems to show that when an algorithm of two-sided geometry needs to test a point against a line, more often than not it needs to take a different course of action for each of the three possible outcomes.

2.2. The two sides of an hyperplane

Let's now examine some special cases of the relative orientation function $\diamond$ in more detail. If a is a point and h is a hyperplane, we say that a *is on the positive side of* h if and only if $a \diamond h = +1$. (To avoid ambiguity, I will not use the names "left side" and "right side" unless a is a point and b is a line of $\mathbf{T}_2$.)

The order of the two operands matters if and only if the space has odd dimension; in that case, to correctly identify the positive side we must put the point on the left side of the $\diamond$ operator.

2.3. Two points on a line

In particular, on the two-sided line $\mathbf{T}_1$ the operation $p \diamond q$ tests whether the points p, q form a positive or a negative simplex of Υ, that is, whether q is ahead of or behind p on Υ_1. In the spherical model, this tests whether the shortest arc from p to q (in the spherical model) is counterclockwise ($p \diamond q = +1$) or clockwise ($p \diamond q = -1$). The test returns $\mathbf{0}$ if $p = q$ or $p = \neg q$. The positive side of a point (hyperplane) q is therefore the half-line *ending* at q, i.e. the arc from $\neg q$ to q.

2.4. The two sides of a plane

A plane π divides $\mathbf{T}_3$ in two half-spaces. By definition, p is on the *positive side* of π if $p \vee \pi = \mathbf{T}_3$, and on the *negative side* if $p \vee \pi = \neg \mathbf{T}_3$. Let $(q; r; s)$ be a positive simplex of π; according to this definition, p is on the positive side of π if and only if $(p; q; r; s)$ is a positive tetrahedron of $\mathbf{T}_3$. See figure 3.

What do the two sides of a plane look like in the straight model? Recall that in the straight model of $\mathbf{T}_3$ a finite plane π is represented by two copies π^+, π^- of some Euclidean plane of $\mathbf{R}^3$, one in each range. Let L^+ and R^+ be the two half-spaces of $\mathbf{R}^3$ determined by π^+ on the front range, and L^-, R^- their antipodes on the back range. The two sides of the plane π then are $L^+ \cup R^-$ and $R^+ \cup L^-$, plus the appropriate points at infinity. See figure 4.

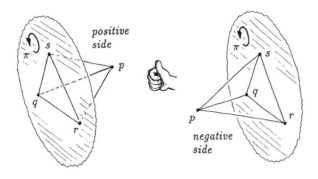

Figure 3. Testing a point against a plane.

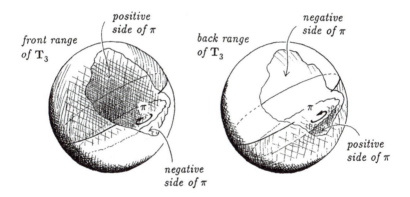

Figure 4. The two sides of a plane in $\mathbf{T}_3$.

Intuitively, the positive side of a plane π in $\mathbf{T}_3$ is the side from which the circular arrow of π seems to turn *clockwise*. Alternatively, we are in the positive side of π if, when we move towards any point q on π by the shortest route, the direction of travel $p \to q$ and the circular arrow of π at q are like the thumb and fingers of the right hand.

Observe that since planes and points have odd rank, $p \vee \pi = \neg(\pi \vee p)$. Therefore, the order of the join above is very important: to test whether a point is on the positive side of a plane, we join the point to the plane, not vice-versa.

2.5. The hyperplane at infinity

The set of all points at infinity of $\mathbf{T}_\nu$ (in the straight model) forms an unoriented hyperplane. I will denote by Ω_ν (or just Ω) that hyperplane, oriented so that the front origin O is on its positive side; that is, $O \diamond \Omega = +1$.

In the case of $\mathbf{T}_3$, the orientation of Ω is what we would obtain by taking the unit 2-sphere of the front range, oriented counterclockwise as seem from the outside, and expanding it to infinite radius. Therefore, for an observer at the origin of the front range, the "big circular arrow in the sky" will be turning *clockwise*.

2.6. Two lines in three-space

A less obvious use of the $\diamond$ function is in testing the relative orientation of two lines l, m in $\mathbf{T}_3$. The test $l \diamond m$ is positive if and only if a positive pair $(p; q)$ on l followed by a positive pair $(r; s)$ on m form a positive tetrahedron $(p; q; r; s)$ of $\mathbf{T}_3$. Intuitively, this means that m "turns around" l according to the right-handed rule. See figure 5. The test is zero if and only if the two lines intersect.

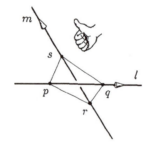

Figure 5. Two positively oriented lines of $\mathbf{T}_3$.

Note that l turns around m the same way that m turns around l. This is to be expected, since the join of two lines (rank 2) is commutative.

3. The separation theorem

An important axiom of Euclidean geometry states that if two points of the plane are distinct, they can be separated by a straight line. This axiom has a counterpart in oriented projective geometry: if two points p, q of $\mathbf{T}_2$ are distinct, there is a line l such that $p \diamond l \neq q \diamond l$. More generally, we have

Theorem 1. *Two flats a, b with the same rank are distinct if and only if there is some flat x in Υ with complementary rank such that $a \diamond x = -(b \diamond x)$.*

PROOF: The "if" is trivial, so let's prove the "only if" part. Let a, b be distinct flats with same rank k. Consider first the case where a and b are disjoint, and $a \vee b = \Upsilon$. That means $\text{rank}(\Upsilon) = 2k$. If $a = \Lambda$, then b must be $-\Lambda$, and $x = \Lambda$ will satisfy the theorem. Otherwise, let $(a^0; \ldots a^k)$ and $(b^0; \ldots b^k)$ be representative simplices of a and b. Construct the simplex $(x^1; \ldots x^\nu)$ where x^i (viewed as a vector of $\mathbf{R}^n$) is $\sigma_i a^i + b^i$, for some coefficients σ_i to be determined.

Now observe that the $2k \times 2k$ matrices M and N which relate the two simplices $(a^0; \ldots a^\kappa; x^0; \ldots x^\kappa)$ and $(b^0; \ldots b^\kappa; x^0; \ldots x^\kappa)$ to $(a^0; \ldots a^\kappa; b^0; \ldots b^\kappa)$ are

$$
M = \begin{pmatrix} 1 & & & & 0 \\ & \ddots & & & \\ 0 & & 1 & & \\ \sigma_0 & & 0 & 1 & \\ 0 & & & \ddots & \\ 0 & & \sigma_\kappa & 0 & 1 \end{pmatrix}
\qquad
N = \begin{pmatrix} 0 & & 0 & 1 & & 0 \\ & \ddots & & & \ddots & \\ 0 & & 0 & 0 & & 1 \\ \sigma_0 & & 0 & 1 & & 0 \\ & \ddots & & & \ddots & \\ 0 & & \sigma_\kappa & 0 & & 1 \end{pmatrix}
$$

Since the determinant of M is $+1$, independently of the σ_i, we have $a \diamond x = +1$. On the other hand, the determinant of N is $(-1)^{k^2} \sigma_0 \cdots \sigma_\kappa$, which can be set to -1 by a suitable choice of the σ_i. In that case we have $b \diamond x = -1$, and $a \diamond x = -(b \diamond x)$.

Now let a and b be arbitrary flats of Υ with same rank k. Let c be their intersection, arbitrarily oriented, and let u, v be right complements of c in a and b, so that $c \vee u = a$ and $c \vee v = b$. Obviously, we must have $u \neq v$, and $\text{rank}(u) = \text{rank}(v) = m$ for some $m \leq k$. Since v is disjoint from $a \cap b$ and contained in b, it is disjoint from a; therefore, $c \vee u \vee v$ is well-defined. Let d be such that $c \vee u \vee v \vee d = \Upsilon$. By the discussion above, there is a flat y of rank m in $u \vee v$ such that $u \vee y = u \vee v = \neg(v \vee y)$. Let x be $y \vee d$; we have

$$a \vee x = c \vee u \vee v \vee d = c \vee u \vee v \vee d = \Upsilon$$
$$b \vee x = c \vee v \vee y \vee d = c \vee \neg(u \vee v) \vee d = \neg\Upsilon$$

Therefore, $a \diamond x = -(b \diamond x)$.

QED.

It follows immediately that

Theorem 2. *A flat a of $\mathbf{T}_\nu$ is uniquely characterized by the sign-valued function $x \mapsto a \diamond x$ from $\mathcal{F}_{\bar{r}}$ into $\{-1, 0, +1\}$, where $\bar{r} = \text{corank}(a)$.*

Note that this result holds even if $a \in \{\Lambda, \neg\Lambda, \Upsilon, \neg\Upsilon\}$, and can be extended to the case where a is $\mathbf{0}^k$, for all k.

4. The coefficients of a hyperplane

In the vector space model, a hyperplane h of $\mathbf{T}_\nu$ is represented by an oriented linear subspace H of $\mathbf{R}^n$ of dimension $n - 1$. For any point $p = [x]$ of $\mathbf{T}_\nu$, the predicate $p \diamond h$ tests on which side of H the vector x lies. If y is any vector of $\mathbf{R}^n$ orthogonal to the space H and directed into its positive side, then testing $p \diamond h$ is equivalent to testing whether the projection of x onto the one-dimensional subspace generated by y has the same sign as y. In other words, $p \diamond h$ tests the sign of the dot product of x and y:

$$p \diamond h = \text{sign}(x_0 y_0 + x_1 y_1 + \cdots + x_\nu y_\nu) \tag{2}$$

From this formula and from the separation theorem it follows that the hyperplane h is uniquely determined by the vector y. The coordinates of y are called the *homogeneous coefficients* of the hyperplane h. I will use $\langle y \rangle$ to denote the hyperplane whose homogeneous coefficients are the coordinates of the vector y. As in the case of homogeneous coordinates of points, it is obvious from equation (2) and from the separation theorem that $\langle y \rangle = \langle z \rangle$ if and only if $y = \alpha z$ for some positive real number α.

It is often convenient to view y as a column vector $y = (y^0; .. y^\nu)$, so that formula (2) can be written as a matrix product

$$[x] \diamond \langle y \rangle = \text{sign}(x_0 y^0 + \cdots + x_\nu y^\nu) = \text{sign}\left((x_0, .. x_\nu) \cdot \begin{pmatrix} y^0 \\ \vdots \\ y^\nu \end{pmatrix} \right)$$

Again, note that in spaces of odd dimension the order of the arguments of $\diamond$ is important, since

$$\langle y \rangle \diamond [x] = (-1)^\nu \left([x] \diamond \langle y \rangle \right) = (-1)^\nu \, \text{sign}(x_0 y^0 + \cdots + x_\nu y^\nu)$$

Chapter 8
Projective maps

The idea of projective maps grew out of the perspective rendering techniques developed by Renaissance artists. The perspective projection consists of extending a line from each point of the scene to the observer's eye, and marking the point where that line intersects the picture plane. See figure 1.

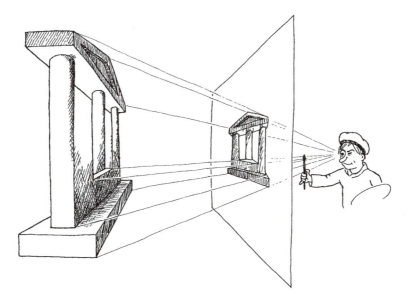

Figure 1. The perspective projection.

As figure 2 shows, even flat objects appear greatly distorted when viewed in perspective. The projection does not preserve many common geometric properties — angles, distances, areas (or their ratios), parallelism, perpendicularity, congruence, and so forth. Indeed, by moving the viewer and the projection screen appropriately, we can make the image of any convex quadrilateral on the floor to match any convex quadrilateral drawn on the screen. Informally, the two-dimensional projective maps are all possible mappings from the floor plane to the screen plane that can be realized in this way.

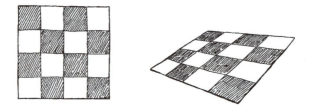

Figure 2. A chessboard, and a perspective view of it.

The perspective projections *do* preserve some attributes of the image, however: they always take straight lines to straight lines. Therefore, they also preserve the incidence relations between points and lines, and all geometric properties that can be defined in terms of incidence. For example, if three points are colinear, or three lines are concurrent, they will remain so in projection (provided we assume that two parallel lines meet at an improper point.) This property implies, for instance, that no perspective view of a quadrilateral can be a circle or a pentagon.

Intuitively, a *projective map* or *collineation* is a generalization of the perspective projection map: it is a function from one projective space into another that takes straight lines to straight lines, and therefore preserves the essential geometric structure of its domain. In this chapter, I will formally define the concept of projective map within the two-sided framework.

1. Formal definition

The standard approach to this subject is to take the line-preserving property as the definition of projective map, and prove from it that every projective map of $\mathbf{T}_\nu$ (for $\nu \geq 2$) corresponds in the analytic model to a linear transformation of $\mathbf{R}^n$. For technical and expository reasons, however, I will reverse this path: I will start with an analytic definition of projective maps as linear transformations of $\mathbf{R}^n$, and prove the line-preserving property that th eline-preserving propety follows from it.

To make formulas more readable, I will use xF as a synonym of $F(x)$, the image of an element x by a function F. I will also denote by FG the composition of functions F and G, applied in that order; that is, $xFG = (xF)G = G(F(x))$. I will write I_A, or just I, for the identity map on a set A, and $\overleftarrow{F}$ for the inverse of a (one-to-one) function F.

Definition 1. Let S and T be two flats of $\mathbf{T}_\nu$, represented in the vector space model by subspaces U and V. A function M from the points of S to those of T is a *projective map* if it takes positive simplices of S to positive simplices of T, and there is a linear map M from U to V such that, for all $u \in U$,

$$M([u]) = [\mathsf{M}(u)]$$

We can say that the linear map M of definition 1 *induces* the projective map M, and denote that by $[\![\mathsf{M}]\!] = M$.

Note that a projective map from S to T is also a projective map from $\neg S$ to $\neg T$, but not from S to $\neg T$. In other words, a projective map does not impose a specific orientation on either its domain or its range, but establishes a connection between the two orientations. A projective map of a flat S to itself is also called an *orientation-preserving map of S*, while a map from S to $\neg S$ is an *orientation-reversing* one.

Of particular interest are the projective maps of $\mathbf{T}_\nu$ to itself, which are induced by linear maps of $\mathbf{R}^n$ to itself. By definition, any such map must preserve the sign of every ν-dimensional proper simplex, which means the inducing linear map must take positive bases of $\mathbf{R}^n$ to positive bases. As we know from linear algebra, this happens if and only if the coefficient matrix of the map has positive determinant. If the marix has negative determinant, the projective map is orientation-reversing, that is, takes $\mathbf{T}_\nu$ to $\neg \mathbf{T}_\nu$.

I will denote by $\mathcal{M}_\nu$ the set of all projective maps from $\mathbf{T}_\nu$ to itself.

1.1. Image of flats

Although a projective map $M = [\![\mathsf{M}]\!]$ is defined as a mapping from points to points, it extends naturally to a mapping from flats to flats by the equations

$$M(\Lambda) = \Lambda$$
$$M(\neg\Lambda) = \neg\Lambda \tag{1}$$
$$M([u^0;\ldots u^\kappa]) = [\mathsf{M}(u^0);\ldots \mathsf{M}(u^\kappa)]$$

for any tuple $u^0;\ldots u^\kappa$ of linearly independent vectors in the domain U of M.

In order for equation (1) to make sense, we have to show that the value of $M([u^0; ..u^\kappa])$ does not depend on the basis $u^0; ..u^\kappa$, but only on the linear space generated by it. In other words, we have to show that

$$[u^0; ..u^\kappa] = [v^0; ..v^\kappa] \Rightarrow [M(u^0); ..M(u^\kappa)] = [M(v^0); ..M(v^\kappa)] \qquad (2)$$

The left-hand side of equation (2) says that the ordered bases $u^0; ..u^\kappa$ and $v^0; ..v^\kappa$ are equivalent, that is, there is a matrix of coefficients $A = (a_j^i \; : \; i,j \in 0..\kappa)$ with positive determinant such that $u^i = \sum_j a_j^i v^j$. Since M is linear, it follows that $M(u^i) = \sum_j a_j^i M(v^j)$, which implies the right-hand-side of equation (2). This result shows that definition 1 is not ambiguous. Note that the requirement in definition 1 that M takes positive simplices of S to positive simplices of T guarantees that the vectors $M(u^0); ..M(u^\kappa)$ in equation (1) are linearly independent.

2. Examples

An example of projective map id the function from $\mathbf{T}_2$ to itself induced by the linear transformation $(w, x, y) \mapsto (w, \; x+w, \; y+w/2)$ of $\mathbf{R}^3$. In the straight model, this map is simply a translation of the front and back ranges by the vector $(1, 1/2)$. Figure 3 shows the effect of this map in the spherical model.

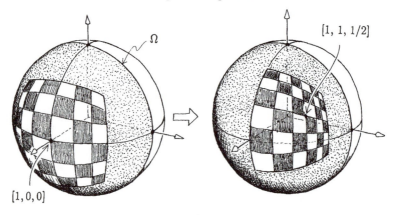

Figure 3. A translation.

Note that every point on the line at infinity Ω remains fixed under this map.

Another example is the map $(w, x, y) \mapsto (10w, 4x - 3y, 3x + 4y)$. The induced projective map performs a rotation around the origin by an angle $\theta = \arctan(3/4)$, combined with a reduction by a factor of $\sqrt{3^2 + 4^2}/10 = 1/2$. See figure 4.

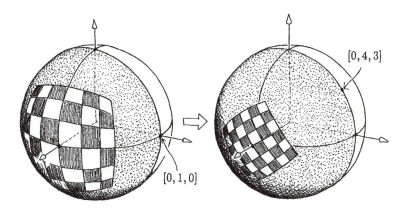

Figure 4. A similarity transformation.

The two examples seen so far map the line at infinity Ω onto itself. It is easy to see that such maps are precisely those that correspond in the straight model to affine transformations of the front and back ranges. A projective map that doesn't belong to this class is the one induced by $(w, x, y) \mapsto (w - x, x, y)$. In the straight model, this map brings the line Ω to the vertical line through $x = -1$. See figure 5.

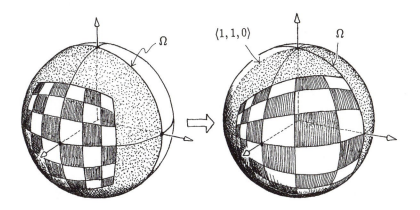

Figure 5. A perspective map.

Its effect can be described as a perspective projection between two copies of $\mathbf{T}_2$ properly positioned in $\mathbf{T}_3$. Note that parallel lines are mapped to lines that converge on the line $x = -1$. Note also that a portion of the back range is mapped onto the front range, and vice-versa.

3. Properties of projective maps

3.1. Continuity

Every projective map $M = [\![M]\!]$ is continuous; this statement follows directly from the continuity of the linear map M, and from the fact that $[u]$ is a continuous function of u. Any projective map M has a functional inverse $\overleftarrow{M} = [\![\overleftarrow{M}]\!]$, and is therefore one-to-one and onto. We have thus proved the following result:

Theorem 1. *Every projective map from a flat S to a flat T is a topological homeomorphism between the point sets of S and T.*

3.2. Group properties

Projective maps are obviously closed under composition. The restriction of a projective map M to a flat subset Z of its domain is a projective map from Z to ZM. For any flat S, the identity function on the points of S is a projective map from S to itself. From these properties, we conclude:

Theorem 2. *For any flat S, the projective maps of S to itself form a group under composition.*

3.3. Meet and join

It follows immediately from (1) that, for every projective map M and any two flats a, b in its domain,

$$(\neg a)M = \neg(aM)$$

and

$$(a \vee b)M = (aM) \vee (bM)$$

Note that a projective map M from a flat S to a flat T takes by definition the universe of S to that of T (including orientation). Since the meet operation is defined in terms of join and the universe, we also have

$$(a \wedge_S b)M = (aM) \wedge_T (bM)$$

From these results we conclude that a projective map from S to T is an isomorphism between the *projective structures* of S and T, which consists of all notions which can be defined in terms of points, flats, orientation reversal, join, and meet.

4. The matrix of a map

A projective map M from $\mathbf{T}_\mu$ into $\mathbf{T}_\nu$ is induced by a linear map M of $\mathbf{R}^m$ into $\mathbf{R}^n$. Such a map can be represented by an $m \times n$ matrix of coefficients

$$
\begin{pmatrix}
m^0_0 & \cdots & m^0_\nu \\
\vdots & \ddots & \vdots \\
m^\mu_0 & \cdots & m^\mu_\nu
\end{pmatrix}
$$

with the convention that the image of a point $x = (x_0, .. x_\nu)$ is

$$
(xM)_j = \sum_i x_i m^i_j \tag{3}
$$

If we view an element of $\mathbf{R}^n$ as a $1 \times n$ (i.e., row-like) matrix, we can express formula (3) as the matrix product

$$
(x_0, .. x_\nu)M = (x_0, .. x_\nu) \cdot
\begin{pmatrix}
m^0_0 & \cdots & m^0_\nu \\
\vdots & \ddots & \vdots \\
m^\nu_0 & \cdots & m^\nu_\nu
\end{pmatrix} \tag{4}
$$

Of course, formulas (3) and (4) describe also the induced map M, since by definition $[x]M = [xM]$. In other words, to compute the image of a point we post-multiply its homogeneous coordinate vector by the coefficient matrix.

A remark on notation: some authors prefer to view the elements of $\mathbf{R}^n$ as *column* vectors, and describe a linear map as *pre*-multiplication by the transpose of the matrix above. I adopted the present convention for consistency with the postfix notation xM and the left-to-right composition rule.

4.1. Examples

Here are some examples of maps from $\mathbf{T}_\nu$ to itself in matrix form, and their effect (described in terms of the straight model):

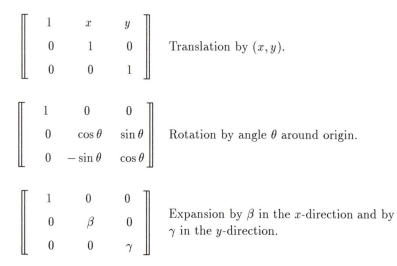

$$\left[\begin{array}{ccc} 1 & x & y \\ 0 & 1 & 0 \\ 0 & 0 & 1 \end{array}\right]$$ Translation by (x, y).

$$\left[\begin{array}{ccc} 1 & 0 & 0 \\ 0 & \cos\theta & \sin\theta \\ 0 & -\sin\theta & \cos\theta \end{array}\right]$$ Rotation by angle θ around origin.

$$\left[\begin{array}{ccc} 1 & 0 & 0 \\ 0 & \beta & 0 \\ 0 & 0 & \gamma \end{array}\right]$$ Expansion by β in the x-direction and by γ in the y-direction.

4.2. Equivalence

Note that the same projective map may be induced by many distinct linear maps. The theorem below characterizes these maps:

Theorem 3. *Two linear maps K, L induce the same projective map if and only if $K = \alpha L$ for some $\alpha > 0$.*

PROOF: The "if" part is a trivial consequence of the definition. As for the "only if" part, assume K and L induce the same map M. Then K and L must have the same domain U and same range V (two linear subspaces of $\mathbf{R}^n$). The composite map $F = K\overleftarrow{L}$ must induce the identity projective map: for any $u \in U$, we must have $[uF] = [u]$.

In particular, let $(u^0; .. u^\kappa)$ be a basis of U. For any two distinct elements u, v of the basis we must have $[uF] = [u]$, $[vF] = [v]$, and $[(u + v)F] = [u + v]$. This means $uF = \alpha u$, $vF = \beta v$, and $(u + v)F = \gamma(u + v)$ for some $\alpha, \beta, \gamma > 0$. Since F is linear, we have also $(u + v)F = (uF) + (vF) = \alpha u + \beta v$, and therefore

$$\alpha u + \beta v = \gamma(u + v) = \gamma u + \gamma v$$

Since u and v are independent, we must have $\alpha = \beta$; and since this is true for all pairs u, v, there is a single $\alpha > 0$ such that $(u^i)F = \alpha u^i$ for all i. We conclude that F is a positive multiple $\alpha \cdot I_U$ of the identity map on U, and therefore

$$K = (K\overleftarrow{L})L = FL = \alpha \cdot I_U L = \alpha \cdot L$$

QED.

From theorem 3 it follows that two projective maps from $\mathbf{T}_\mu$ into $\mathbf{T}_\nu$ are equal if and only if their matrices are positive multiples of each other. Therefore, we can identify a projective map M with the class of all matrices M that induce it. Accordingly, I will write

$$\left[\!\!\left[\begin{array}{ccc} m_0^0 & \cdots & m_\nu^0 \\ \vdots & \ddots & \vdots \\ m_0^\mu & \cdots & m_\nu^\mu \end{array} \right]\!\!\right]$$

to denote the projective map generated by that matrix.

4.3. Image of a hyperplane

Geometric computations deal with hyperplanes almost as often as they do with points, so we it is important to study the effect of a projective map M on the homogeneous coefficients of a hyperplane h.

First let's assume M is a positive (orientation-preserving) map. Such a map must preserve the relative orientation of points and hyperplanes, and therefore

$$p \diamond h = (pM) \diamond (hM) \qquad \text{for all points } p \text{ of } \mathbf{T}_\nu. \tag{5}$$

In fact, this property completely characterizes the hyperplane hM.

Now let $p = [u]$ and $h = \langle v \rangle$. If we view u as a row matrix, and v as a column one, then $p \diamond h$ is the sign of the matrix product $u \cdot v$ (which is a scalar). If M is the matrix of map M, and $\overline{M}$ is its inverse, we have also the obvious identity

$$u \cdot v = (u \cdot M) \cdot (\overline{M} \cdot v) \tag{6}$$

The term $u \cdot M$ is a row vector, the coordinates of the point pM. The term $\overline{M} \cdot v$ is a column vector, the coefficients of some hyperplane g. Equation (6) says therefore

$$p \diamond h = (pM) \diamond g \qquad \text{for all } p \in \mathbf{T}_\nu.$$

Comparing this statement with equation (5), we conclude that $g = hM$. That is, we obtain the coefficients of hM by the matrix product $\overline{M} \cdot v$.

This conclusion is correct only as long as the determinant $|M|$ of M's matrix is positive. If $|M|$ is negative, then in equation (5) we must compute the second $\diamond$ relative to the space $\neg \mathbf{T}_\nu$; that is, equation (5) becomes $p \diamond h = \neg((pM) \diamond (hM))$. On the other hand, equation (6) is not affected, so the end result is $hM = [\![-\overline{M} \cdot v]\!]$.

We can condense both cases into the single formula $\text{sign}(|M|)\cdot\overleftarrow{M}\cdot v$. Moreover, since positive factors are irrelevant, we can replace $\text{sign}(|M|)\cdot\overleftarrow{M}$ by just $|M|\cdot\overleftarrow{M}$. This matrix is the so-called *adjoint* or *adjugate* of the linear map M. Therefore,

Theorem 4. *For any hyperplane $\langle v\rangle$ and any non-degenerate projective map $M = [\![M]\!]$, we have $\langle v\rangle M = \langle \overline{M}\cdot v\rangle$, where $\overline{M}$ is the adjoint linear map $|M|\cdot\overleftarrow{M}$ of M.*

Chapter 9
General two-sided spaces

The preceding chapters defined the canonical two-sided spaces $\mathbf{T}_\nu$ of arbitrary dimension ν. It is now time to define general two-sided spaces in a more abstract way. Among other things, this will allow us to view any flat of rank k in $\mathbf{T}_\nu$ as a copy of the canonical space $\mathbf{T}_\kappa$, and will prepare the ground for the discussion of duality in chapter 10.

1. Formal definition

Definition 1. The *canonical oriented projective space of dimension* ν is the quadruple $\mathbf{T}_\nu = (\mathcal{F}_\nu, \mathcal{M}_\nu, \vee_\nu, \wedge_\nu)$.

Definition 2. An *oriented projective space of dimension* ν is a quadruple $S = (\mathcal{F}_S, \mathcal{M}_S, \vee_S, \wedge_S)$, isomorphic to $\mathbf{T}_\nu = (\mathcal{F}_\nu, \mathcal{M}_\nu, \vee_\nu, \wedge_\nu)$.

By "isomorphic" we mean that (i) $\mathcal{M}_S$ must be a group of bijections from $\mathcal{F}_S$ to itself, (ii) $\vee_S$ and $\wedge_S$ must be binary operations on $\mathcal{F}_S$, and (iii) there must be a one-to-one mapping φ from $\mathcal{F}_S$ to $\mathcal{F}_\nu$ such that

$$\mathcal{M}_S = \{\, \overleftarrow{\varphi} M \varphi \ : \ M \in \mathcal{M}_\nu \,\} \tag{1}$$

$$(a\varphi) \vee_\nu (b\varphi) = (a \vee_S b)\varphi \tag{2}$$

$$(a\varphi) \wedge_\nu (b\varphi) = (a \wedge_S b)\varphi \tag{3}$$

hold for all a, b in $\mathcal{F}_S$. Informally, condition (1) says that $\mathcal{M}_S$ should act on the elements of $\mathcal{F}_S$ in the same way that $\mathcal{M}_\nu$ acts on the corresponding elements of $\mathcal{F}_\nu$. In particular, for every bijection $M \in \mathcal{M}_S$ there should be a projective map M^φ of $\mathbf{T}_\nu$ such that $M^\varphi = \overleftarrow{\varphi} M \varphi$, that is,

$$(sM)\varphi = (s\varphi)M^\varphi \quad \text{for all } s \in \mathcal{F}_S.$$

As we saw in chapter 8, every projective map of $\mathbf{T}_\nu$ to itself is also an isomorphism of $\mathbf{T}_\nu$ to itself (an *automorphism* of $\mathbf{T}_\nu$). In fact, it is possible to prove that the only automorphisms of a projective space S are the projective maps of S to S.

1.1. Geometric operations

Additional geometric operations in a generic two-sided space S can be defined in terms of its join and meet operations. Equations (2) and (3) imply that the operations defined this way will coincide with the corresponding operations in $\mathbf{T}_\nu$, mapped through any isomorphism φ of S to $\mathbf{T}_\nu$. For example, the rank of any element $a \in \mathcal{F}_S$, defined as the number of points we have to join to get a, is the same as the rank in $\mathbf{T}_\nu$ of the flat $a\varphi$. The vacuum Λ_S of S (the neutral element of $\vee_S$) will always be $\Lambda\overleftarrow{\varphi}$, and the universe Υ_S of S (the neutral element of $\wedge_S$) will be $\Upsilon_\nu\overleftarrow{\varphi}$. Similarly, the operations of orientation reversal (join with Λ_S) and the relative orientation predicate (which compares the result of $\vee_S$ with Υ_S) satisfy

$$\neg_S\, a = (\neg\, a\varphi)\overleftarrow{\varphi} \tag{4}$$

$$a \diamond_S b = (a\varphi) \diamond (b\varphi) \tag{5}$$

where $\neg$, $\vee$, $\wedge$, and $\diamond$ denote the standard operations of $\mathbf{T}_\nu$.

2. Subspaces

We have already observed informally that a κ-dimensional flat of $\mathbf{T}_\nu$ looks pretty much like a copy of $\mathbf{T}_\kappa$. We can now state this more precisely:

Theorem 1. *Let*

s *be a κ-dimensional flat of $\mathbf{T}_\nu$,*

$\mathcal{F}_s$ *be the set of all flats of $\mathbf{T}_\nu$ contained in s,*

$\mathcal{M}_s$ *be the set of all projective maps from s to s,*

$\vee_s$ *be the join of $\mathbf{T}_\nu$, restricted to subflats of s, and*

$\wedge_s$ *be the meet operation relative to s.*

Then $(\mathcal{F}_s, \mathcal{M}_s, \vee_s, \wedge_s)$ is an oriented projective space isomorphic to $\mathbf{T}_\kappa$.

PROOF: Let $([u^0]; .. [u^\kappa])$ be any positive simplex of s; in other words, $(u^0; .. u^\kappa)$ is a positive basis of the oriented linear subspace of $\mathbf{R}^n$ that represents s in the vector space model. Now consider the linear map E from $\mathbf{R}^k$ into $\mathbf{R}^n$ given by the matrix product

$$(x^0, .. x^\kappa) \; \mapsto \; (x^0, .. x^\kappa) \cdot \begin{pmatrix} u_0^0 & \cdots & \cdots & u_\nu^0 \\ \vdots & & & \vdots \\ u^\kappa & \cdots & \cdots & u_\nu^\kappa \end{pmatrix}$$

Let η be the map from $\mathbf{T}_\kappa$ into $\mathbf{T}_\nu$ defined by $[x]\eta = [xE]$ for all points $[x]$ of $\mathbf{T}_\kappa$, and extended to arbitrary flats of $\mathbf{T}_\kappa$ by the formula

$$(\Lambda_\kappa)\eta = \Lambda_\nu$$
$$(\neg\Lambda_\kappa)\eta = \neg\Lambda_\nu$$
$$[u^0; .. u^\kappa]\eta = [E(u^0); .. E(u^\kappa)]$$

These are essentially the same formulas we used for projective maps in chapter 8, and by the same arguments used there we can prove that η is well-defined. It is also easy to check that η is an isomorphism from $\mathbf{T}_\kappa$ to $(\mathcal{F}_s, \mathcal{M}_s, \vee_s, \wedge_s)$. QED.

I will call $(\mathcal{F}_s, \mathcal{M}_s, \vee_s, \wedge_s)$ the *subspace of* $\mathbf{T}_\nu$ *determined by* s. Not surprisingly, the orientation reversal operation in this subspace is that of $\mathbf{T}_\nu$, restricted to $\mathcal{F}_s$. The relative orientation test $a \diamond_s b$ checks whether $a \vee b$ is s or $\neg s$.

Note that the flats a and $\neg a$ determine different subspaces (they have different universes), even though $\mathcal{F}_a = \mathcal{F}_{\neg a}$ and $\mathcal{M}_a = \mathcal{M}_{\neg a}$. In particular, the subspace of $\mathbf{T}_\nu$ determined by $\neg \Upsilon_\nu$ is the space $\neg\mathbf{T}_\nu = (\mathcal{F}_\nu, \mathcal{M}_\nu, \vee, \bar\wedge)$, where $x \bar\wedge y = \neg(x \wedge y)$. In this subspace the relative orientation predicate $\bar\diamond$ is such that $x \bar\diamond y = \neg(x \diamond y)$.

2.1. The canonical inclusion map

It is often useful to identify the space $\mathbf{T}_\kappa$ (for all $\kappa < \nu$) with the flat of $\mathbf{T}_\nu$ generated by the first m points of the standard simplex of $\mathbf{T}_\nu$. This flat consists of all points of $\mathbf{T}_\nu$ whose last $n - m$ coordinates are zero. The *canonical embedding* of $\mathbf{T}_\kappa$ in $\mathbf{T}_\nu$ is the function η that appends $n - k$ zeros to the coordinates of every point in $\mathbf{T}_\kappa$. Informally, η takes every point of $\mathbf{T}_\kappa$ to the "same" point of $\mathbf{T}_\nu$.

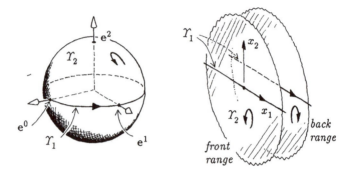

Figure 1. The canonical embedding of $\mathbf{T}_1$ in $\mathbf{T}_2$.

For example, the canonical embedding of $\mathbf{T}_1$ in $\mathbf{T}_2$ (in the spherical model) maps the unit circle $\mathbf{S}_1$ to the great circle determined on $\mathbf{S}_2$ by the plane of the first two coordinate axes. In the straight model, this great circle corresponds to the x-axis of the front and back ranges. See figure 1. Analytically, point $[w, x]$ of the two-sided line is mapped to point $[w, x, 0]$ of the two-sided plane.

2.2. Bundles

We can generalize even further the notion of subspaces induced by flats as follows. Let s, t be two flats of $\mathbf{T}_\nu$, such that $s \supseteq t$. The *bundle* $\mathcal{F}_{s:t}$ determined by s and t consists of all flats that are contained in flat s and contain flat t. I will denote by $\mathcal{M}_{s:t}$ the set of all projective maps from s to s that take the flat t to itself, with each map restricted to the flats that contain t. I will also define the *join relative to t* of two flats a, b in this bundle, denoted by $\vee_t$, by the formula

$$(t \vee x) \vee_t (t \vee y) = t \vee x \vee y \tag{6}$$

I will let the reader verify that this definition is consistent, and that t is its the neutral element of $\vee_t$. We can prove now the following result:

Theorem 2.   *For any two flats $s \supseteq t$, the quadruple $(\mathcal{F}_{s:t}, \mathcal{M}_{s:t}, \vee_t, \wedge_s)$ is a projective space of rank* $\mathrm{rank}(s) - \mathrm{rank}(t)$.

To prove this theorem, let c be any subflat of s that is a right complement of t in s (that is, $t \vee c = s$). Then consider the mapping φ from $\mathcal{F}_c$ to $\mathcal{F}_{s:t}$ defined by $x\varphi = t \vee x$ for every $x \in \mathcal{F}_c$. The inverse of this map takes every flat y in the bundle to the flat $y \wedge_s c$ of $\mathcal{F}_s$, where $\wedge_s$ is the meet operation relative to the flat s. It is easy to check that $\overleftarrow{\varphi}$ is an isomorphism from $(\mathcal{F}_{s:t}, \mathcal{M}_{s:t}, \vee_t, \wedge_s)$ to the subspace induced by c.

For example, consider the bundle $\mathcal{K}$ of all flats that are contained in the plane $\pi = \Upsilon_2$ and contain the front origin $o = [1, 0, 0]$. See figure 2. The

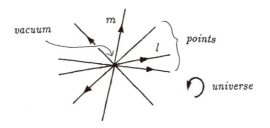

Figure 2. The bundle of all lines through the origin of $\mathbf{T}_2$.

flats of this bundle are the point o and its antipode, all the lines passing through o (with coefficients of the form $\langle 0, X, Y \rangle$), and the two planes π and $\neg\pi$.

The join operation $\vee_o$ of this bundle satisfies $o \vee_o x = x \vee_o o = x$ and $(\neg o) \vee_o \dot{x} = x \vee_o (\neg o) = \neg x$ for all $x \in \mathcal{K}$. Also, for any two lines $l, m \in \mathcal{K}$, the join $l \vee_o m$ is either Υ_2 or $\neg\Upsilon_2$, according to whether the angle from l to m at o is positive or negative. Finally, for all other flats $a, b \in \mathcal{K}$, we have $a \vee_o b = \mathbf{0}^{\text{rank}(a)+\text{rank}(b)-1}$. The meet operation $\wedge_\pi$ of this bundle is the standard meet operation of $\mathbf{T}_2$, and $\mathcal{M}_{\pi:o}$ is the set of all projective maps of $\mathbf{T}_2$, each restricted to the flats in $\mathcal{K}$. One of the many isomorphisms from $\mathbf{T}_1$ to this bundle is the map $\Lambda \mapsto o$, $\neg\Lambda \mapsto \neg o$, $[x, y] \mapsto \langle 0, -y, x \rangle$, $\Upsilon_1 \mapsto \pi$, and $\neg\Upsilon_1 \mapsto \neg\pi$.

Other examples of bundles are the (one-dimensional) space of all planes of $\mathbf{T}_3$ containing a given line, or the (two-dimensional) space of all lines and planes of $\mathbf{T}_3$ containing a given point. Note that if we take $t = \Lambda$ we get simply the subspace determined by s. If we take $t = \neg\Lambda$, we get a space whose join operation $\bar{\vee}$ satisfies $x \bar{\vee} y = \neg(x \vee y)$.

We should note here that including both the projective maps and the operations of join and meet in the definition of a projective space is generally an overkill. More precisely, in spaces of dimension 2 or more a projective map can be defined as a bijection of $\mathcal{F}$ that commutes with the join operation. However, in one-dimensional spaces this requirement is too weak, and is satisfied by many functions that cannot be expressed as linear maps of the homogeneous coordinates. Accepting those functions as projective maps would be extremely inconvenient. For instance, our current definition ensures that every projective map defined on a line a can be extended to a projective map for any plane containing a; but this property would not be true if projective maps were defined in terms of join alone. For similar reasons, we cannot define join and meet of a two-sided space S in terms of the projective maps $\mathcal{M}_S$ alone.

Once the join operation is defined, we can define the meet operation in terms of the universe, or the universe in terms of meet. I have chosen the second alternative because it treats meet and join more symmetrically, and so smoothes the way to our discussion of duality in chapter 10.

Chapter 10
Duality

The reader may have observed that meet and join have very similar properties. In fact, most of the formulas we have seen so far occurred in pairs, where one member of the pair can be transformed into the other by exchanging meet with join, points with hyperplanes, rank with co-rank, vacuum with universe, and so forth. Compare for example the formulas

$$a \vee \Lambda = a \qquad\qquad a \wedge \Upsilon = a$$
$$a \vee (\neg b) = \neg(a \vee b) \qquad a \wedge (\neg b) = \neg(a \wedge b)$$
$$b \vee a = \neg^{rs}(a \vee b) \qquad b \wedge a = \neg^{\bar{r}\bar{s}}(a \wedge b)$$

where

$$r = \operatorname{rank}(a) \qquad \bar{r} = \operatorname{corank}(a)$$
$$s = \operatorname{rank}(b) \qquad \bar{s} = \operatorname{corank}(b)$$

This phenomenon is known as the principle of *projective duality*. Its unoriented version is one of the most important ideas of classical projective geometry. In this chapter we will see that it holds in oriented geometry as well.

1. Duomorphisms

The duality between meet and join follows from a rather subtle result, whose proof will be given in section 3:

Theorem 1. *The quadruple* $\mathbf{T}_\nu^* = (\mathcal{F}, \mathcal{M}, \wedge, \vee)$ *is a projective space isomorphic to* $\mathbf{T}_\nu = (\mathcal{F}, \mathcal{M}, \vee, \wedge)$.

In other words, there exists a one-to-one mapping η from $\mathcal{F}$ to $\mathcal{F}$ that satisfies $\overleftarrow{\eta} M \eta \in \mathcal{M}$ for all $M \in \mathcal{M}$, and

$$(a\eta) \wedge (b\eta) = (a \vee b)\eta$$
$$(a\eta) \vee (b\eta) = (a \wedge b)\eta$$

for all $a, b \in \mathcal{F}$. That is, join in the space $\mathbf{T}_\nu^*$ is the same as meet in $\mathbf{T}_\nu$, and vice-versa.

83

As discussed in chapter 9, the isomorphism η must also satisfy

$$\Lambda\eta = \Upsilon$$
$$\Upsilon\eta = \Lambda$$
$$\neg(a\eta) = (\neg a)\eta$$
$$(a\eta) \diamond (b\eta) = a \diamond b$$
$$\text{rank}(a\eta) = \text{corank}(a)$$

The space $\mathbf{T}^*_\nu$ is the *dual space* of $\mathbf{T}_\nu$, and any isomorphism between the two spaces (such as η above) is a *duomorphism* of $\mathbf{T}_\nu$.

1.1. Formal duality

The duomorphisms of $\mathbf{T}_\nu$ provide a solid foundation for the duality principle. Let $\mathcal{E}$ be a formula or assertion about the flats of $\mathbf{T}_\nu$, with no free variables, involving logical connectives and set operations, plus the symbols Λ, Υ, $\mathcal{F}$, $\mathcal{M}$, $\neg$, $\vee$, and $\wedge$, and any other operations that can be defined in terms of those. We construct the *formal dual* $\mathcal{E}^*$ of $\mathcal{E}$ by exchanging every occurrence of Λ with Υ, $\wedge$ with $\vee$, and recursively any derived concept with its formal dual. That includes swapping rank with co-rank, the word "point" with "hyperplane," the predicate $a \subseteq b$ (for flats) with $a \supseteq b$, and so on.

For example, the assertion "point x is on the segment pq" can be written as "$\text{rank}(p) = \text{rank}(q) = \text{rank}(x) = 1$ and $p \vee x = x \vee q = p \vee q \neq \mathbf{0}$." The dual of this formula is "$\text{corank}(p) = \text{corank}(q) = \text{corank}(x) = 1$ and $p \wedge x = x \wedge q = p \wedge q \neq \mathbf{0}$," which in $\mathbf{T}_2$ means the line x is concurrent with p and q, and its direction lies in the shorter angle between the directions of p and q.

As another example, consider predicate $a \diamond b = +1$. To construct its dual, we first rewrite it in terms of join, obtaining $a \vee b = \Upsilon$. The formal dual of this expression is $a \wedge b = \Lambda$. By the definition of $\wedge$, this is the same as $a \vee b = \Upsilon$. We conclude that $\diamond$ is its own dual. One can easily verify that the same is true of the orientation reversal operation ($\neg$).

Meta-theorem 2. *If $\mathcal{T}$ is a theorem of oriented projective geometry, then its formal dual $\mathcal{T}^*$ is also a theorem.*

The proof of this theorem is a straightforward exercise in formal logic, which we will not prove here since it falls somewhat outside the scope of this work.

2. The polar complement

It is now time to prove theorem 1. To do so I will exhibit one particular duomorphism for $\mathbf{T}_\nu$, the *polar complement* function.

2.1. Polar flats

I will say that two points of $\mathbf{T}_\nu$ are *polar* if they are represented by orthogonal vectors in the spherical model. In general, two flats a, b are *polar*, denoted by $a \perp b$, if every point of one flat is polar to every point of the other.

2.2. Polar complement

Recall that, according to theorem 1 of chapter 5, for every flat a in Υ there is some *complementary flat* b such that $a \vee b = \Upsilon$ (and therefore $a \wedge b = \Lambda$, and $a \diamond b = +1$). In general, the flat b is not unique; for example, the complement of a point p in $\mathbf{T}_2$ can be any line of $\mathbf{T}_2$ that leaves p on its left side. However, we can use the polarity predicate above to make the complement unique:

Definition 1. The *right polar complement* of a flat a is the flat $a^\vdash$ such that

$$a \perp a^\vdash$$
$$a \diamond a^\vdash = +1 \tag{1}$$

It is not hard to see that $a^\vdash$ always exists, is unique, is a continuous function of a, is disjoint from a, and satisfies $\mathrm{rank}(a^\vdash) = \mathrm{corank}(a)$. Symmetrically, the *left polar complement* $\dashv$ is defined by

$$a^\dashv \perp a$$
$$a^\dashv \diamond a = +1 \tag{2}$$

As usual, it is convenient to define also $(\mathbf{0}^k)^\vdash = (\mathbf{0}^k)^\dashv = \mathbf{0}^{n-k}$ for all k. The names "right complement" and "left complement" do not refer to the relative positions of those flats in $\mathbf{T}_\nu$, but to the order of a, $a^\vdash$, and $a^\dashv$ in formulas (1–2): for the result to be $+1$, the right complement must go on the right side of the "$\diamond$", and the left complement must go on the left side. The symbols $\vdash$ and $\dashv$ are supposed to make this rule easier to remember.

It follows immediately from the definition that

$$(a^\vdash)^\dashv = a = (a^\dashv)^\vdash$$

That is, $\vdash$ and $\dashv$ are inverses of each other.

We also have

$$(\neg a) \vee (\neg(a^\vdash)) = a \vee a^\vdash = \Upsilon$$
$$(\neg a^\dashv) \vee (\neg(a)) = a^\dashv \vee a = \Upsilon$$

which implies

$$(\neg a)^\vdash = \neg(a^\vdash)$$
$$(\neg a)^\dashv = \neg(a^\dashv)$$

Let rank$(a) = r$ and rank$(\Upsilon) = n$; from the properties of $\diamond$,

$$a^\vdash \diamond a = \neg^{r(n-r)} \left(a \diamond a^\vdash \right)$$

which leads to

$$a^\vdash = \neg^{r(n-r)} a^\dashv \tag{3}$$

and

$$(a^\vdash)^\vdash = (a^\dashv)^\dashv = \neg^{r(n-r)} a$$

Observe that if n is odd, then the product $r(n - r)$ is even for all r. Therefore, in spaces of odd rank (even dimension) $\dashv$ and $\vdash$ are the same function. This is the case, for example, in the plane $\mathbf{T}_2$. To emphasize this fact, I will in those spaces use the same symbol $a^\perp$ for both $a^\vdash$ and $a^\dashv$. On the other hand, in spaces of even rank (odd dimension), like $\mathbf{T}_3$, we have $a^\vdash = a^\dashv$ or $a^\vdash = \neg a^\dashv$, depending on whether the rank of a is even or odd. So, for example, the left and right complements of a point (rank 1) or a plane (rank 3) are opposite, whereas those of a line are the same.

2.3. Polar complements in the two-sided plane

Let's consider some examples. In the two-sided plane the two polar complements are the same function, $a^\vdash = a^\vdash = a^\perp$. In the spherical model, this function takes every oriented great circle of $\mathbf{S}_2$ to the apex of its left hemisphere, and, conversely, every point of $\mathbf{S}_2$ to the oriented great circle whose left hemisphere has that point at the apex. See figure 1(a). In the straight model, the image $l^\perp$ of a finite line l passing at distance $d > 0$ from the origin O is the point p such that the vector Op is perpendicular to l, directed away from l, and with length $1/d$. See figure 1(b). The point $l^\perp$ will be on the front range if and only if the line is directed counterclockwise as seen from the front origin, i.e. if the origin lies on the positive side of l.

(a)   (b)

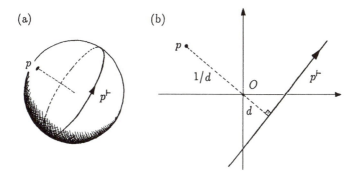

Figure 1. Polar complement in $\mathbf{T}_2$.

If the line l passes through the origin (i.e., $d = 0$), then $l^\perp$ is the point at infinity in the direction $90°$ counterclockwise from that of l. Conversely, if $l = \Omega$, then $l^\perp = O$ (the origin of the front range), and if $l = \neg\Omega$ then $l^\perp = \neg O$ (the origin of the back range). Observe how in all cases these definitions put $l^\perp$ on the left (positive) side of l.

2.4. Polar complements in three-space

Let us now consider the case of a point p in three-dimensional space $\mathbf{T}_3$. If p is on the front range and at distance $d > 0$ from the origin O, its polar complement $p^\perp$ is a plane perpendicular to the vector Op and at distance $1/d$ from O in the direction opposite to that of p. See figure 2.

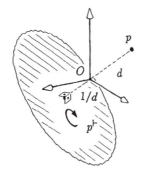

Figure 2. The polar complement of a point in $\mathbf{T}_3$.

If p is the front origin O, its polar complement is one of the two planes at infinity. If p is on the back range, the same applies with $\neg O$ substituted for O. If p is infinite, then the plane passes through O and is perpendicular to the direction from O towards p. Since $p \diamond p^{\llcorner} = +1$, $p^{\llcorner}$ is oriented so that p is on its positive side; that is, the circular arrow of $p^{\llcorner}$ turns *clockwise* as seen from p. In particular, $O^{\llcorner} = \Omega$. Since p has odd rank (1) and odd co-rank (3), we have $p^{\llcorner} = \neg p^{\lrcorner}$.

Consider now a finite line l of $\mathbf{T}_3$ whose point p closest to the origin is at distance $d > 0$ from it. The right polar complement $l^{\llcorner}$ is a line skew to but perpendicular to l. The point q of $l^{\llcorner}$ that is closest to O is at distance $1/d$ from O, in the direction opposite to Op. See figure 3.

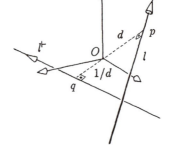

Figure 3. The polar complement of a line of $\mathbf{T}_3$.

If l is at infinity, $l^{\llcorner}$ passes through the origin and is perpendicular to all planes containing l. The orientation of $l^{\llcorner}$ is given by the right-hand rule. Since lines have even rank, their left and right polar complements coincide: $l^{\llcorner\llcorner} = l^{\lrcorner\lrcorner} = l$.

2.5. Polar complement in the analytic model

In the vector space model, the right polar complement of a point $[x]$ is a hyperplane of $\mathbf{T}_\nu$, represented by a $(n-1)$-dimensional linear subspace of $\mathbf{R}^n$ that is orthogonal to the vector x. The homogeneous coefficients of that hyperplane are the coordinates of a properly oriented vector orthogonal to that linear subspace, that is, either x or $-x$ (or any multiple thereof). In other words, we have $[x] = \langle \sigma x^{\mathrm{tr}} \rangle$, where x^{tr} is x viewed as a column vector, and $\sigma \in \{+1, -1\}$.

We can figure out the correct sign from the defining equation $a \diamond a^{\vdash} = +1$ and $[x] \diamond \langle z \rangle = \text{sign}(x_0 z^0 + \cdots + x_\nu z^\nu)$, which imply

$$[x]^{\vdash} = \langle x^{\text{tr}} \rangle$$

It follows immediately that

$$\langle z \rangle^{\vdash} = \neg^\nu [z^{\text{tr}}] \tag{4}$$

and, symmetrically,

$$[x]^{\dashv} = \neg^\nu \langle x^{\text{tr}} \rangle \tag{5}$$

$$\langle z \rangle^{\dashv} = [z^{\text{tr}}] \tag{6}$$

3. Polar complements as duomorphisms

Let's then prove that $\vdash$ and $\dashv$ are indeed duomorphisms. Most of the proof is contained in the next two lemmas:

Lemma 3. *For any projective map M of $\mathbf{T}_\nu$, the functions $\overleftarrow{\vdash} M \vdash$ and $\overleftarrow{\dashv} M \dashv$ are projective maps of $\mathbf{T}_\nu$.*

PROOF: Let M be a linear map of $\mathbf{R}^n$ that induces M. Equation (5) says that, for any point $[x]$ of $\mathbf{T}_\nu$, $[x]^{\dashv} = \neg^\nu \langle x^{\text{tr}} \rangle$. From theorem 4 of chapter 8 we know that

$$\langle x^{\text{tr}} \rangle M = \langle \overline{M} \cdot x^{\text{tr}} \rangle$$

where $\overline{M} = |M| \cdot \overleftarrow{M}$ is the adjoint of the matrix of M. Finally, from equation (4) we get

$$\langle \overline{M} \cdot x^{\text{tr}} \rangle^{\vdash} = \neg^\nu [(\overline{M} \cdot x^{\text{tr}})^{\text{tr}}]$$

Collecting all pieces together, we get

$$[x]^{\overleftarrow{\vdash}} M \vdash \; = \; [x]^{\dashv} M \vdash \; = \; [x \cdot \overline{M}^{\text{tr}}]$$

that is, $\overleftarrow{\vdash} M \vdash$ is the projective map $[\![\overline{M}^{\text{tr}}]\!]$, whose matrix is the *untransposed* cofactor matrix of M.

To conclude the proof, observe that, for every point x,

$$x(\overleftarrow{\dashv} M \dashv) \; = \; \left(\neg^\nu \right)^2 x(\overleftarrow{\vdash} M \vdash)) \; = \; \overleftarrow{\vdash} M \vdash$$

which shows $\overleftarrow{\dashv} M \dashv$ is the same projective map as $\overleftarrow{\vdash} M \vdash$.
QED.

Lemma 4. *The polar complement functions satisfy*

$$(a \vee b)^\vdash = a^\vdash \wedge b^\vdash \qquad\qquad (a \vee b)^\dashv = a^\dashv \wedge b^\dashv$$
$$(a \wedge b)^\vdash = a^\vdash \vee b^\vdash \qquad\qquad (a \wedge b)^\dashv = a^\dashv \vee b^\dashv$$

for all flats (or null objects) a, b.

PROOF: Let's prove first that $(a \vee b)^\vdash = a^\vdash \wedge b^\vdash$. Let $c = (a \vee b)^\vdash$, $r = \text{rank}(a)$, and $s = \text{rank}(b)$, so that $\text{rank}(c) = n - r - s \geq 0$. (Recall that for the join to be defined, we must have $r + s \leq n$).

From the definition of $\vdash$, we get $(a \vee b) \diamond (a \vee b)^\vdash = +1$, that is,

$$a \vee b \vee c = \Upsilon \tag{7}$$

which implies

$$a^\vdash = b \vee c \tag{8}$$

Also, from equation (7) and the commutativity laws of join we have

$$b \vee c \vee a = \neg^{r(n-r)} \Upsilon \tag{9}$$

which implies

$$b^\vdash = \neg^{r(n-r)} (c \vee a) \tag{10}$$

From this statement and from equation (8), we get

$$a^\vdash \wedge b^\vdash = (b \vee c) \wedge \left(\neg^{r(n-r)} (c \vee a) \right)$$
$$= \neg^{r(n-r)} \left((b \vee c) \wedge (c \vee a) \right)$$

From equation (9) and the definition of $\wedge$, we have

$$(b \vee c) \wedge (c \vee a) = \neg^{r(n-r)} c$$

and therefore

$$a^\vdash \wedge b^\vdash = \left(\neg^{r(n-r)} \right)^2 c = (a \vee b)^\vdash \tag{11}$$

as we proposed to show. For the left complement, from equation (3) we get

$$a^\dashv = \neg^{r(n-r)} a^\vdash$$
$$b^\dashv = \neg^{s(n-s)} b^\vdash$$
$$(a \vee b)^\dashv = \neg^{(r+s)(n-r-s)} (a \vee b)^\vdash$$

which, plugged into equation (11), gives

$$(a \vee b)^{\dashv} = \neg^{(r+s)(n-r-s)+r(n-r)+s(n-s)} \left(a^{\dashv} \wedge b^{\dashv}\right)$$
$$= \neg^{2rn+2sn-2r^2-2rs-2s^2} \left(a^{\dashv} \wedge b^{\dashv}\right) \tag{12}$$
$$= a^{\dashv} \wedge b^{\dashv}$$

Applying $\dashv$ to both sides of (11), and performing the substitutions $a^{\vdash} \mapsto a$, $b^{\vdash} \mapsto b$ (which are valid, since $\dashv$ and $\vdash$ are one-to-one and onto) we get

$$(a \wedge b)^{\dashv} = a^{\dashv} \vee b^{\dashv}$$

In the same way, from equation (12) we get $(a \wedge b)^{\vdash} = a^{\vdash} \vee b^{\vdash}$. This concludes the proof.

QED.

The main result of this chapter then follows trivially from lemmas 3 and 4:

Theorem 5. *The function $\vdash$ is an isomorphism between the projective spaces $(\mathcal{F}, \mathcal{M}, \wedge, \vee)$ and $(\mathcal{F}, \mathcal{M}, \vee, \wedge)$.*

To conclude this section, observe that, for every flat a of rank r,

$$a(\overleftarrow{\dashv M \dashv}) = \left(\neg^{r(n-r)}\right)^2 a(\overleftarrow{\vdash M \vdash}) = \overleftarrow{\vdash M \vdash}$$

which shows that $\dashv$ too is a duomorphism of $\mathbf{T}_\nu$.

4. Relative polar complements

We can generalize the definition of polar complement to arbitrary subspaces of $\mathbf{T}_\nu$ by using an arbitrary flat f in lieu of the universe. That is, for any flat $x \subseteq f$ we define the *right polar complement of x relative to f* as the flat $x \rceil f$ satisfying

$$x \perp (x \rceil f)$$
$$x \vee (x \rceil f) = f$$

Symmetrically, the *left polar complement of x relative to f* satisfies

$$(f \lceil x) \perp x$$
$$(f \lceil x) \vee x = f$$

In particular, $x^{\vdash} = x \rceil \Upsilon$, $x^{\dashv} = \Upsilon \lceil x$. Conversely, we have

$$x \rceil f = f \wedge x^{\vdash}$$
$$f \lceil x = x^{\dashv} \wedge f$$

These results are special cases of the following theorem:

Theorem 6. *For any flats* $x \subseteq f \subseteq g,$

$$x \rceil f = f \wedge (x \rceil g)$$
$$f \lceil x = (g \lceil x) \wedge f$$

PROOF: First, observe that $f \wedge (x \rceil g)$) is contained in f and is polar to x (since $x \rceil g$ is polar to x). From $x \vee (x \rceil f) = f$ we get $x \vee (x \rceil f) \vee (f \rceil g) = \Upsilon$, hence

$$(x \rceil f) \vee (f \rceil g) = (x \rceil g)$$

Therefore
$$f \wedge (x \rceil g) = [x \vee (x \rceil f)] \wedge [(x \rceil f) \vee (f \rceil g)]$$
$$= x \rceil f$$

QED.

Lemma 4 generalizes to relative complements through the equations

$$(a \vee b) \rceil f = (a \rceil f) \wedge (b \rceil f)$$
$$f \lceil (a \vee b) = (f \lceil a) \wedge (f \lceil b)$$
$$(a \wedge b) \rceil f = (a \rceil f) \vee (b \rceil f)$$
$$f \lceil (a \wedge b) = (f \lceil a) \vee (f \lceil b)$$

There are no simple general formulas for $a \rceil (f \vee g)$, $a \rceil (f \wedge g)$, or their $\lceil$ equivalents.

5. General duomorphisms

In general, a *duomorphism* is an isomorphism between a two-sided space $S = (\mathcal{F}_S, \mathcal{M}_S, \vee_S, \wedge_S)$ and its *dual space* $S^* = (\mathcal{F}_S, \mathcal{M}_S, \wedge_S, \vee_S)$.

It is easy to check that the composition of a duomorphism η and a projective map (in either order) is also a duomorphism. In particular, the composition of the polar complement $\dashv$ and any projective map of $\mathbf{T}_\nu$ to itself is a duomorphism of $\mathbf{T}_\nu$.

Conversely, if η and φ are duomorphisms from S to S^*, then the composition $\eta\overleftarrow{\varphi}$ is obviously an isomorphism of S to itself, and therefore a projective map of S. It follows that

Theorem 7. *Every duomorphism of a space S can be written as the product of a fixed duomorphism of S and some projective map of S.*

In fact, if $\text{rank}(S) = m$, we can always write a duomorphism η of S as $\eta = M \vdash \overleftarrow{N}$, where M and N are projective maps from S to $\mathbf{T}_\mu$, and $\vdash$ is the right polar complement in $\mathbf{T}_\mu$. Actually, we can choose one of M and N arbitrarily, with the other being a function of η and the chosen map.

In particular, any duomorphism of $\mathbf{T}_\nu$ is the composition of the right polar complement $\vdash$ and a suitable projective map of $\mathbf{T}_\nu$ (or vice-versa). For example, the left polar complement $\dashv$ of $\mathbf{T}_\nu$ is the composition of $\vdash$ and the projective map

$$x \mapsto \neg^{n-1} x \quad \text{for every point } x.$$

Note how this map is the identity for spaces of even dimension (meaning $\dashv$ and $\vdash$ are the same function), and is the antipodal map $x \mapsto \neg x$ for spaces of odd dimension.

6. The power of duality

Duality is an extremely powerful tool. For one thing, it greatly reduces the number of theorems that have to be proved, since every proof automatically establishes the correctness of a theorem and its dual. Moreover, we can choose among the two theorems the one whose proof is easier to visualize, so we may end up doing much less than half the work.

Duality is equally valuable from a computational point of view, since it allows every geometrical algorithm to do the work of two. Thus, a subroutine that computes $a \vee b$ can be used to compute $a \wedge b$, by the formula $\left(a^\vdash \vee b^\vdash\right)^\dashv$. As we shall see, given the proper representation, $\dashv$ and $\vdash$ can be computed at negligible cost. Duality thus may cut the size of a geometric library (and of its documentation) by almost one half. Similar savings apply to higher-level algorithms; for example, an algorithm that computes the convex hull of n points can also be used to find the intersection of n half-spaces.

Chapter 11
Generalized projective maps

The one-to-one projective maps defined in chapter 8 are important because they are exactly the projective isomorphisms; that is, they are the only functions from flats to flats that preserve straight lines and orientations (and threfore also join, meet, and all derived concepts). In this chapter we will examine a larger class of functions from flats to flats that are almost as good as the projective maps: these functions are not one-to-one, but still preserve the geometric structure of space to some extent.

1. Projective functions

Recall that a projective map M was defined as a correspondence between two flats S, T that is determined by an invertible linear transformation M acting on the homogeneous coordinates of points of S. We can generalize this definition by dropping the requirement that the linear map M be invertible.

More precisely, let S and T be two flats of $\mathbf{T}_\nu$, not necessarily of the same rank, represented in the vector space model by subspaces U and V. Let M be an arbitrary linear function from U into V, and consider the function $M = [\![M]\!]$ from the points of S into those of T that is induced by M according to

$$[x]M = [xM] \quad \text{for all } x \in \mathbf{R}^n.$$

I will call any function M defined this way a *projective function* from S into T. It is easy to show that two linear maps M, N induce the same projective function if and only if $M = \alpha N$ for some positive real α.

As we know from linear algebra, the map M is many-to-one if and only if it takes some non-zero vector of U to the zero vector $(0, .. 0)$. When this happens, the induced function M will take some valid point of S to the null object $\mathbf{0}$. The set of such points is a flat subset of S, corresponding to the non-zero vectors in the null space (kernel) of the map M. By analogy, I will call that subset of S the *null space* of the function M, and denote it by $Null(M)$.

I will denote by $Range(M)$ the flat set consisting of all the points of T that are images of some point in S. Note that this flat set need not span the whole of T.

95

Indeed, as we know from linear algebra, the rank of $Range(M)$ is the rank of S minus the rank of $Null(M)$.

The natural extension of the function M to arbitrary subflats of S is

$$\Lambda M = \Lambda$$
$$(\neg\Lambda)M = \neg\Lambda$$
$$[u^0; .. u^\kappa]M = [(u^0)M; .. (u^\kappa)M]$$

If the linear map M is not one-to-one, then the images of k independent vectors $u^0; .. u^\kappa$ from U may not be independent vectors of V. For the images to be independent, the linear subspace $<u^0; .. u^\kappa>$ or U must be disjoint from the null space of M; that is, the flat subset of S spanned by the simplex $([u^0]; .. [u^\kappa])$ must be disjoint from the flat set $Null(M)$.

We conclude that the image of a flat a by a projective function M is well-defined if and only if the flats a and $Null(M)$ are independent.

When a and $Null(M)$ intersect, the image of a is best defined as the null object $\mathbf{0}$ with the same rank as a. We may be tempted to define aM in this case as the unoriented flat set $\{ xM : x \in a \}$, which has rank strictly less than that of a; however, this extension would not be of much help in practice, and would make many other formulas and theorems needlessly complicated. This question is closely related to that of assigning meaning to $a \vee b$ when a and b are not disjoint, and most arguments relevant to the latter apply also to the former.

Note that, unlike a projective map, the projective function M is not required to preserve the sign of every representative simplex of its domain S. In fact, if M is not a true projective map, this requirement does not even make sense. If the inducing linear map M is not onto, then the image of a maximum-rank simplex of S has less than maximum rank in T, and its orientation is not defined. Conversely, if the map M is many-to-one, then any maximum-rank simplex of S will be mapped to a degenerate simplex in T.

1.1. The perspective projection

An important example of a projective function is the perspective projection of $\mathbf{T}_3$ that we mentioned in chapter 8. Let π be the picture plane, and p the position of the observer. The perspective projection will map an arbitrary point x on the object being rendered to the point

$$F(x) = (p \vee x) \wedge \pi \tag{1}$$

of the image plane. See figure 1. This projection is clearly not one-to-one, since it maps all the points in the ray ox to the same point of π.

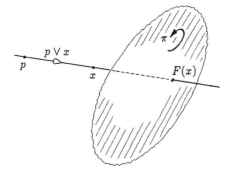

Figure 1. Perspective projection.

Formula (1) is well defined as long as the point p is not on the plane π, and x is neither p nor $\neg p$. According to our previous conventions, we should let $F(p)$ and $F(\neg p)$ be the null object $\mathbf{0}$ of rank 1. When p is on π, then formula (1) specifies a degenerate map that takes every point x to $\neg p$, p, or $\mathbf{0}$, depending on whether $x \diamond \pi$ is positive, negative, or zero. In the sequel, we will assume this is not the case.

Formula (1) can be used also to give the perspective projection of an arbitrary flat x of $\mathbf{T}_3$. From the formula it is obvious that $F(x)$ is well defined as long as the flat x is not incident to the point p. For example, the projection of a line of $\mathbf{T}_3$ is a well-defined line of π, unless the line is seen end-on. When the flat x is incident to p, the formula defines $F(x)$ as the null object $\mathbf{0}$ with same rank as x.

For perspective rendering it is convenient to orient π so that p is on its positive side. This choice implies $F(x) = x$ for all x on π; that is, F is idempotent, as a projection map should be. When p is on the negative side of π we obviously have $F(x) = \neg x$ for all flats x contained in π.

In typical perspective rendering applications, π is a proper plane (not contained in Ω), and p is a point on the front range. Note that all points on the plane parallel to π and passing through p will project to a point at infinity on π. In the Cartesian framework those points must be handled as special cases, but of course in the projective framework they are just like any other point.

1.2. General and polar projections

Note that formula (1) can be extended to use any pair of complementary flats in lieu of p and π. That is, if a, b are subflats of a flat S with $a \vee b = S$, we define the *projection of S onto b from a* as the mapping

$$F(x) = (a \vee x) \wedge b$$

where the meet is computed relative to S. For example, if a and b are two skew lines in $\mathbf{T}_3$, then F will be a map taking every point of $\mathbf{T}_3$ that is not on line a to a point on line b. The points of $\mathbf{T}_3$ with the same image y are precisely those on the plane $a \vee y$, except for those on a itself (which are mapped to $\mathbf{0}$). See figure 2.

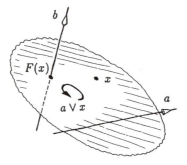

Figure 2. Projecting $\mathbf{T}_3$ onto a line.

Why is F a projective function? Let A, B, and U be the linear subspaces representing a, b, and S, respectively, in the vector space model. Note that A and B are complementary subspaces of U. Now consider the linear map M that decomposes each vector x of U into its A and B components, and returns the latter. That is, M projects every vector x onto the space B in a direction parallel to A. It is easy to check that this linear map M induces precisely the function F defined above.

A particularly interesting case is when a and b are polar complementary flats of $\mathbf{T}_\nu$, with $a = b^\perp$. In that case I call F the *polar projection* of $\mathbf{T}_\nu$ on b. In the vector space model, F corresponds to orthogonal projection of $\mathbf{R}^n$ onto the linear subspace B of $\mathbf{R}^n$ that represents b.

1.3. Properties of projective functions

It follows immediately from the definition that projective functions are closed under composition. Also, if F is a projective function with domain S, then the restriction of F to any subflat X of S is also projective.

Like a projective map, a projective function M has the property that $(a \vee b)M = (aM) \vee (bM)$, except that M is allowed to produce the zero flat even when the argument is nonzero. The analogous formula for $\wedge$ is not valid when stated in this way, essentially because the universe of the domain flat S is not mapped to that of the range flat T. However, if X is any flat of S disjoint from the null space of M, then the image $Y = XM$ is a well-defined flat of T, and the restriction of M to X

is a true projective map from X to XM. In that case, for two flats a, b contained in X we have

$$(a \wedge_X b)M = (aM) \wedge_Y (bM)$$

In particular, we can take X of maximum rank subject to $X \cap Null(M) = \emptyset$, that is, we can let X be any flat of S that is complementary to $Null(M)$. In that case the set Y will cover the actual range of M.

1.4. Natural domain of projective functions

Given a projective function defined on $\mathbf{T}_\nu$ (or any flat thereof) we can use the standard polarity relation to pick a particular flat set X of maximum rank such that the restriction of M to X is non-degenerate. That set is simply the polar complement of the function's null space, $(Null(M))^\perp$. I call that set the *natural domain* of M, and denote it by $Dom(M)$. The restriction of M to $Dom(M)$ is non-degenerate and has the same range as M.

Conversely, any non-degenerate map M from a flat X of $\mathbf{T}_\nu$ to some other flat Y can be extended to a projective function N from the whole $\mathbf{T}_\nu$ onto Y; the extension consists of the polar projection of $\mathbf{T}_\nu$ onto X followed by the given map M. The function N is the *polar extension* of M to $\mathbf{T}_\nu$. In particular, the polar projection of $\mathbf{T}_\nu$ onto a flat X of $\mathbf{T}_\nu$ is the polar extension of the identity map on X.

1.5. The inverse of a projective function

Since a projective function may be many-to-one, it may not have an inverse in the ordinary sense. However, if M is an arbitrary projective function from $\mathbf{T}_\nu$ into $\mathbf{T}_\mu$, we can consider the restriction N of M to its natural domain $Dom(M)$, as described above, take the inverse of N (which is well-defined), and finally take the polar extension this inverse to a projective function from $\mathbf{T}_\mu$ into $\mathbf{T}_\nu$. I will call the resulting map $\overleftarrow{M}$ the *(generalized) inverse* of M.

The matrix of the pseudo-inverse $\overleftarrow{M}$ turns out to be the so-called *generalized least-squares inverse* or *Moore-Penrose inverse* of the coefficient matrix M of M. In matrix terms, the generalized inverse is the unique matrix $\overleftarrow{M}$ such that

$$(M \overleftarrow{M})^{\mathrm{tr}} = M \overleftarrow{M} \qquad M \overleftarrow{M} M = M$$
$$(\overleftarrow{M} M)^{\mathrm{tr}} = \overleftarrow{M} M \qquad \overleftarrow{M} M \overleftarrow{M} = \overleftarrow{M}$$

These conditions essentially say that $[\![M \overleftarrow{M}]\!]$ and $[\![\overleftarrow{M} M]\!]$ are polar projections onto the domain and range of M, respectively.

It is worth noting that the generalized inverse above does not satisfy the equation $xM\overleftarrow{M} = x$, unless $x \in Dom(M)$. For the same reason, we don't have

$\overleftarrow{MN} = \overleftarrow{N}\,\overleftarrow{M}$, unless $Range(M) = Dom(N)$.

1.6. Topological properties

Projective functions that are not one-to-one obviously are not homeomorphisms, but they still preserve some of the topological structure of the projected objects. The two theorems below show that projective functions preserve open and closed sets — most of the time.

Theorem 1. *A projective function F from a space D into a space R maps closed subsets of $D \setminus Null(F)$ to closed subsets of R.*

PROOF: In terms of the spherical model, a projective function F is a linear function F followed by central projection onto the unit sphere. As long as the argument is not in the null space of F, these operations are well-defined and continuous.

Therefore, F is continuous on the set $D^+ = D \setminus Null(F)$, which means F maps closed subsets of D^+ to closed subsets of its range $F(D^+)$. But since $F(D^+)$ is a flat subset of R, which is closed in R, any set that is closed relative to $F(D^+)$ is also closed in R.

QED.

The following theorem is a bit less trivial. Note that the range of F here is the whole R:

Theorem 2. *A projective function F of a space D onto a space R maps open subsets of $D \setminus Null(F)$ to open subsets of R.*

PROOF: Let X be an open subset of $D \setminus Null(F)$. If X is empty, $F(X)$ is empty and the theorem is trivially true. If X is not empty, then $Null(F)$ is a proper subset of D. Let then x be any point of X, and let Y be a subspace of D with maximum rank such that $x \in Y$ and $Y \cap Null(F) = \emptyset$. It is easy to see that Y must be a complementary flat for $Null(F)$, which means F maps Y onto R in a one-to-one fashion.

Since F is continuous outside $Null(F)$, it is a homeomorphism of Y onto R, and maps open sets of the former to open sets of the latter. Since X is open, $X \cap Y$ is open relative to Y, so $F(X \cap Y)$ is open relative to R. Since $x \in X \cap Y \subseteq X$, we also have $F(x) \in F(X \cap Y) \subseteq F(X)$.

This argument shows that every point $F(x)$ of $F(X)$ is contained in some subset of $F(X)$ that is open in R. We conclude that $F(X)$ is an open subset of R.

QED.

2. Computer representation

In chapter 8 we saw that a projective map from $\mathbf{T}_\nu$ to $\mathbf{T}_\nu$ or to $\neg\mathbf{T}_\nu$ can be represented by an $n \times n$ matrix of coefficients

$$\left[\!\!\left[\begin{array}{ccc} m_0^0 & \cdots & m_\nu^0 \\ \vdots & & \vdots \\ m_0^\nu & \cdots & m_\nu^\nu \end{array} \right]\!\!\right] \tag{2}$$

with nonzero determinant. Projective functions have the same representation, except that the matrix may have zero determinant. In fact, we have

Theorem 3. *Every projective function of* $\mathbf{T}_\nu$ *into itself is completely described by an $n \times n$ real matrix. The function is degenerate if and only if the matrix has zero determinant.*

As in the case of projective maps, two matrices determine the same map if and only if they are positive multiples of each other.

2.1. Matrix of a perspective projection

An important example is the perspective projection M defined by $xM = (a \vee x) \wedge b$, where a and b are complementary flats of $\mathbf{T}_\nu$. If we know positive simplices $a^0; \ldots a^\kappa$ for a and $b^0; \ldots b^\mu$ for b, we can compute the matrix of M by the formula $M = \overleftarrow{P}B$, where B and P are the $n \times n$ matrices

$$P = \begin{pmatrix} a_0^0 & \cdots & & \cdots & a_\nu^0 \\ \vdots & & & & \vdots \\ a_0^\kappa & \cdots & & \cdots & a_\nu^\kappa \\ b_0^0 & \cdots & & \cdots & b_\nu^0 \\ \vdots & & & & \vdots \\ b_0^\mu & \cdots & & \cdots & b_\nu^\mu \end{pmatrix} \qquad B = \begin{pmatrix} 0 & \cdots & & \cdots & 0 \\ \vdots & & & & \vdots \\ 0 & \cdots & & \cdots & 0 \\ b_0^0 & \cdots & & \cdots & b_\nu^0 \\ \vdots & & & & \vdots \\ b_0^\mu & \cdots & & \cdots & b_\nu^\mu \end{pmatrix} \tag{3}$$

Briefly, the effect of $\overleftarrow{P}$ is to compute the coordinates of a given vector u of $\mathbf{R}^n$ relative to the basis formed by the vectors $a^0; \ldots a^\kappa; b^0; \ldots b^\mu$. The subsequent multiplication by B throws away the a^i components of u, and collects the b^i components. The correctness of the formula $M = [\!\![\overleftarrow{P}B]\!\!]$ follows readily from this observation.

2.2. Domain, range, and null space

Let M be a projective function with coefficient matrix M, and let $\mathbf{e}^0 = (1, 0, 0, \ldots 0)$, $\mathbf{e}^1 = (0, 1, 0, \ldots 0)$, $\ldots$, $\mathbf{e}^\nu = (0, 0, 0, \ldots 1)$ be the canonical basis of $\mathbf{R}^n$. Observe that row i of matrix M homogeneous coordinates of the image of point $[\mathbf{e}^i]$ through M. In general, the coordinates of the image of any point x of $\mathbf{T}_\mu$ by M will be a linear combination of the rows of M. Therefore,

Theorem 4. *The range of a projective function from $\mathbf{T}_\mu$ into $\mathbf{T}_\nu$ is the the flat set of $\mathbf{T}_\nu$ corresponding to the linear subspace spanned by the rows of its coefficient matrix.*

Observe also that the jth coordinate of xM is the dot product of the coordinate vector of x by the jth column of matrix M. Therefore, a point of $\mathbf{T}_\mu$ is mapped to $\mathbf{0}$ if and only if its coordinate vector is orthogonal to every column of M. In other words, the null space of the linear map with matrix M is the orthogonal complement in $\mathbf{R}^m$ of the subspace spanned by the columns of M. It follows that

Theorem 5. *The natural domain of a projective function from $\mathbf{T}_\mu$ into $\mathbf{T}_\nu$ is the flat set of $\mathbf{T}_\mu$ corresponding to the subspace spanned by the columns of its coefficient matrix.*

As we know from linear algebra, the row and column spaces of a matrix have the same dimension, which is the size of the largest submatrix with non-zero determinant. This number is the *rank* of the map, and coincides with the geometric rank of the flat sets $Range(M)$ and $Dom(M)$. When the rank is equal to the number of rows m, the map is one-to-one; when the rank is equal to the number of columns n, the map ranges over the whole $\mathbf{T}_\nu$.

In light of the above, an $m \times n$ matrix M can be taken to represent either a (possibly degenerate) projective function M from $\mathbf{T}_\mu$ into $\mathbf{T}_\nu$, or a (non-degenerate) projective map N from $Dom(M)$ to $Range(M)$. Fortunately, in most practical situations (for example, when implementing a basic geometric operations library) it is not necessary to worry about this distinction, since mapping a point x through M or N is done by the same formulas and using the same coefficient matrix. Therefore, both operations can be implemented by a single procedure.

In the same vein, we can use a single procedure to compute both the generalized inverse $\overleftarrow{M}$ of M (which is a projective function from $\mathbf{T}_\nu$ into $\mathbf{T}_\mu$) and the plain inverse $\overleftarrow{N}$ of N (a proper map from $Range(M)$ to $Dom(M)$). Algorithms for computing the matrix of $\overleftarrow{M}$ can be found in the numerical analysis literature [10, 18].

2.3. The canonical embedding map

It is often useful to identify the space $\mathbf{T}_\mu$ (for all $\mu < \nu$) with the flat S of $\mathbf{T}_\nu$ generated by the first m points of the standard simplex of $\mathbf{T}_\nu$. This flat consists of all points of $\mathbf{T}_\nu$ whose last $n - m$ coordinates are zero. The *canonical embedding* of $\mathbf{T}_\mu$ into $\mathbf{T}_\nu$ is the function η that takes every point of $\mathbf{T}_\mu$ to the "same" point of S. In analytic terms, η simply appends $n - m$ zeros to the homogeneous coordinates of its argument; its matrix is

$$
\overbrace{\hspace{2em}}^{m} \quad \overbrace{\hspace{2em}}^{n-m}
$$

$$
\begin{bmatrix}
1 & & 0 & 0 \cdots 0 \\
& \ddots & & \vdots \;\; \vdots \\
0 & & 1 & 0 \cdots 0
\end{bmatrix}
\tag{4}
$$

The generalized inverse of η is the projective function $\overleftarrow{\eta}$ that simply throws away the last $n - m$ coordinates of its argument, and is therefore the polar projection of $\mathbf{T}_\nu$ onto S. The matrix of $\overleftarrow{\eta}$ is just the transpose of (4).

2.4. Alternative representations

The coefficient matrix is the most natural representation of a projective function, but not necessarily the most convenient. For one thing, computing the image of a point x under the inverse map $\overleftarrow{M}$ requires that we solve a system of n linear equations on n unknowns, which requires $\Omega(n^3)$ operations; this is a lot more expensive than computing the direct image, which can be done in $O(n^2)$ steps. Computing the image of a hyperplane h given its coefficients is equally expensive.

One way to reduce these costs is to precompute the inverse matrix $\overleftarrow{M}$ (or the adjoint $\bar{M}$), and store it along with the matrix M. Then points and hyperplanes can be mapped equally fast, in $O(n^2)$ time. One disadvantage of this double representation is that it takes twice as much space (32 real numbers instead of 16, in case of $\mathbf{T}_3$), and therefore it takes twice as long to copy and compose with other maps.

2.5. The LU factorization

We can get the best (and worst?) of both approaches by storing the matrix M in some compact factored form such that mapping by both M and $\overleftarrow{M}$ can be performed relatively fast. For example, we know from linear algebra that any $m \times n$ matrix with $m \leq n$ can be factored into the product of a row permutation matrix, an $m \times m$ lower triangular matrix, and an $m \times n$ upper triangular matrix:

$$
M = LU = \begin{pmatrix} l_0^0 & & 0 \\ \vdots & \ddots & \\ l_0^\mu & \cdots & l_\mu^\mu \end{pmatrix} \begin{pmatrix} \overbrace{u_0^0 \;\; \cdots \;\; u_\mu^0}^{m} & \overbrace{u_\nu^0 \;\; \cdots \cdots \;\; u_\nu^0}^{n-m} \\ & \ddots & \vdots \\ 0 & u_m^\mu u \;\; \cdots \cdots \;\; u_\nu^\mu \end{pmatrix}
$$

This *Gaussian LU factorization* can be computed in $O(mn^2)$ time, occupies only $mn + O(m+n)$ words of storage, and still allows us to compute th eimage of apoint in $O(mn)$ time. If the matrix is square ($m = n$), the inverse mapping too can be computed from the factored form at roughly the same cost as the direct one. One drawback of this format is that computing the composition of two maps requires running the Gaussian elimination algorithm, and is therefore a bit more expensive than with the naive format (but only by a constant factor).

2.6. The singular value decomposition

An alternative to the Gaussian LU decomposition that is worth considering is the *singular value decomposition* (SVD). An arbitrary $m \times n$ real matrix M (with $m \le n$) can always be factored as the product of three matrices $U \in \mathbf{R}^{m \times m}$, $\Sigma \in \mathbf{R}^{m \times n}$, and $V \in \mathbf{R}^{n \times n}$ such that U and V are orthogonal, and Σ is all zero except for the elements on the main diagonal:

$$
M = U\Sigma V^{\mathrm{tr}} = \begin{pmatrix} u_0^0 & \cdots & u_\mu^0 \\ \vdots & & \vdots \\ u_0^\mu & \cdots & u_\mu^\mu \end{pmatrix} \begin{pmatrix} \overbrace{\sigma_0 \;\;\;\; 0}^{m} & \overbrace{0 \cdots 0}^{n-m} \\ & \ddots & \vdots \;\;\; \vdots \\ 0 & \sigma_\mu \;\; 0 \cdots 0 \end{pmatrix} \begin{pmatrix} v_0^0 & \cdots & \cdots & v_0^\nu \\ \vdots & & & \vdots \\ \\ v_\nu^0 & \cdots & \cdots & v_\nu^\nu \end{pmatrix}
$$

The case $m > n$ is similar, except that the matrix Σ will have more rows than columns. It is always possible to arrange for U and V to have positive determinant, and for the numbers $|\sigma_i|$ (the *singular values* of the matrix M) to be sorted in non-increasing order. Furthermore, we can arrange for all σ_i to be non-negative, except perhaps for σ_0. Algorithms for computing this decomposition in time roughly $O(mn(m+n))$ are well documented in the numerical analysis literature [10, 18].

Compared to the LU decomposition, the SVD has the advantage of treating domain and range in a more symmetric fashion. In fact the generalized inverse of M is $V \overleftarrow{\Sigma} U^{\mathrm{tr}}$, where $\overleftarrow{\Sigma}$ is the transpose of Σ with every nonzero σ_i replaced by $1/\sigma_i$. Therefore, the SVD allows points to be mapped through both M and its inverse at the same cost, even when the matrix M is non-square or singular.

Note that when $m \geq n$ we only have to store the first m rows of V. More precisely, if the map's range is a flat set of rank r, then only the first r of the σ_i will be non-zero, which means we only have to store the first r columns of U and the first r rows of V. Therefore the SVD can be represented in $r(n+m) + O(m+n)$ floating-point words, while still allowing points to be mapped in $O(r(m+n))$ time. In fact, by representing the matrices U and V as the product of Householder reflections [10, sec. 3.3] it is possible to bring the storage cost down to $r(m+n) - r^2 + O(m+n)$, without increasing the asymptotic cost of point mapping by more than a constant factor. These optimizations make the SVD a reasonable alternative to LU decomposition for general maps, and a definite win for highly degenerate maps.

Most of the properties of the map M that are related to the generalized inverse are readily obtainable from the singular value decomposition of its matrix. For example, the rank of M is the number of non-zero σ_i. The null space of M is the space spanned by the rows of U whose corresponding σ_i are zero. The SVD decomposition has also good numerical properties when computations are carried out in floating point. The main disadvantage of the SVD is that it takes somewhat longer to compute, and it makes map composition substantially harder.

One can imagine many other possible representations for projective maps, based on other factorizations or more exotic schemas. Each representation has its advantages and disadvantages, and its value depends, among other things, on the relative frequency of the various operations for which it is used. At this point, I can only say that determining the "best" representation of projective maps for general use is still an open problem.

Chapter 12
Projective frames

Frames play the same role in projective geometry as bases do in linear algebra, and as coordinate systems do in physics and Cartesian geometry. Informally, a frame is a geometric object that can be used as a reference in order to assign unique and unambiguous numeric coordinates to every point of some space.

An even more important use of frames is in the description of projective maps for input to programs and subroutines. Instead of writing down the transformation matrix, it is generally much easier for the user or programmer to give a pair of frames, and ask for the map that takes one frame to the other. As we shall see, such a map exists (and is unique) if and only if the corresponding parts of the two frames have the same relative orientation.

1. Nature of projective frames

Recall that in the vector space model a projective map between two subspaces R, S of $\mathbf{T}_\nu$ with rank k is a linear map between two k-dimensional linear spaces U, V of $\mathbf{R}^n$. As we know from linear algebra, such a map is completely specified by giving k independent vectors in U and their images in V. We may falsely deduce that a projective map from R to S can be specified by giving k independent points (i.e., a proper simplex) on R and their images on S. Unfortunately, this is not the case.

The difficulty is that a point of S specifies only the direction of a vector of V, and not its length. For example, suppose we want a map of $\mathbf{T}_2$ to itself that takes

$$
\begin{array}{llll}
a = [1,0,0] & \text{to} & p = [1,0,0] \\
b = [0,1,0] & \text{to} & q = [1,5,3] \\
c = [0,0,1] & \text{to} & r = [1,1,6]
\end{array}
$$

An obvious choice is the map

$$
M = \begin{bmatrix} 1 & 0 & 0 \\ 1 & 5 & 3 \\ 1 & 1 & 6 \end{bmatrix}
$$

However, the point $p = [1, 0, 0]$ can also be written as $[2, 0, 0]$, and therefore the map below will also work:

$$N = \begin{bmatrix} 2 & 0 & 0 \\ 1 & 5 & 3 \\ 1 & 1 & 6 \end{bmatrix}$$

Yet the two maps are different, since, for example, $[1, 1, 1]M = [3, 6, 9] = (2, 3)$, whereas $[1, 1, 1]N = [4, 6, 9] = (1.5, 2.25)$. Obviously, there are infinitely many projective maps of $\mathbf{T}_2$ to itself that take the simplex $(a; b; c)$ to $(p; q; r)$.

As we shall see in the following sections, we can resolve this ambiguity by giving one extra point (besides a, b, and c) and its desired image. In general, to define a projective map between two spaces R and S of rank n, it is necessary and sufficient to give $n + 1$ points of R and their images in S. Thus, it seems reasonable to define a projectve frame for a rank-n space as being a set of $n + 1$ points.

However, another way of resolving the ambiguity of the example above is to specify a line l of $\mathbf{T}_2$ and its image under the desired map. More generally, to specify a projective map from R to S we should give n points and one hyperplane of R, and their images in S. Thus, we could also define a projectve frame of a rank-n space as being a set of n points and one hyperplane.

Frames consisting entirely of points may seem to be simpler than mixed bags of points and hyperplanes, but the latter are actually easier to handle than the former. For one thing, the matrix of the map connecting two given frames is easier to compute for mixed frames than for all-point ones. Mixed frames are also the natural choice for specifying affine maps, as we will see in chapter 15.

The moral of this story is that there is no "natural" definition of a projective frame; rather, there are several different types of frame, each with its own merits and drawbacks. Therefore, we must content ourselves with a rather abstract definition of "frame," such as the one given below.

1.1. Arrangements

An *arrangement* is any finite ordered sequence $a = (a^0, a^1, \ldots, a^\mu)$ of flats in some space S. The *type* of such an arrangement is the integer sequence $(r_0, r_1, \ldots r_\mu)$, where $r_i = \mathrm{rank}(a^i)$. The *span* of a is the flat set $Span(a)$ of S with smallest dimension that contains every element of a. The *rank* and the *dimension of a* are by definition those of its span.

For example, a proper κ-dimensional simplex is an arrangement of dimension κ, rank k, and type $(1, \ldots 1)$. Another example is a list of two points p, q and one line l; its type is $(1, 1, 2)$, and its rank can be 2, 3, or 4, depending on the relative positions of p, q, and l.

1.2. Frames

We are interested in arrangements that can be used to specify projective maps between two spaces. We would like to be able to unambiguously specify a projective map from a space R to a space S by giving some arrangement a on R and its desired image b on S.

Of course, the arrangements cannot be arbitrary. For one thing, they must be *projectively similar*: that is, there must exist some projective map M such that $M(a^i) = b^i$ for all i. For example, two proper simplices with the same number of vertices are similar. Moreover, that map must be unique.

Let's say that an arrangement a is *categorical* if every projective map that takes a to itself also takes every point of $Span(a)$ to itself. Categorical arrangements have the required "unique map" property:

Theorem 1. *If a and b are similar categorical arrangements, there is exactly one projective map from $Span(a)$ to $Span(b)$ that takes a to b.*

PROOF: Since a and b are similar, there is some projective map that takes a to b. Its domain is a flat set that contains a, and therefore $Span(a)$. Let M be the restriction of that map to $Span(a)$. The range of M is a flat set containing b, and therefore it contains $Span(b)$. Moreover, $\overleftarrow{M}(Span(b))$ is a flat set containing a; it follows that the range of M is exactly $Span(b)$.

Now let G be any projective map from $Span(a)$ to $Span(b)$ that takes a to b. The composition $\overleftarrow{M}G$ takes a to itself, and therefore must take every point of $Span(a)$ to itself. It follows that $\overleftarrow{G} = \overleftarrow{M}$, and therefore $M = G$; that is, the map M is unique.

QED.

Theorem 1 justifies the following definition:

Definition 1. A *frame* for a flat set S is a categorical arrangement whose span is exactly the set S.

Recall that any two oriented projective spaces with same dimension are related by some projective map. Therfore, if a is a frame for $\mathbf{T}_\kappa$, then every κ-dimensional two-sided space S contains arrangements that similar to a. It is easy to check that all these arrangements are frames for S.

Obviously, two arrangements can be similar only if they have the same type, but this condition is not sufficient. For example, a degenerate simplex has the same type as a proper one, but the two are not similar. The general problem of characterizing similar and categorical arrangements is relatively well-studied but rather hard; see for example the paper by Crapo and Ryan [8].

2. Classification of frames

Definition 1 is quite abstract, to such an extent that it does not even fix the size or type of the frame. In particular, if $a = (a^0, \ldots a^\mu)$ is a frame, and f is any flat obtained from $a^0, \ldots a^\mu$ by join and relative complement operations, then the arrangement $(a^0, \ldots a^\mu, f)$ is also a frame, with same span as a.

Clearly, there are many "flavors" (similarity classes) of projective frames, even for a fixed flat set S. What is more, there is no single "flavor" that is convenient for all applications. Fortunately, for the purpose of implementing basic geometric libraries, it is enough to consider only frames of two simple flavors, *point frames* and *mixed frames*.

2.1. Point frames

A *point frame* for a flat set S of rank k is an arrangement of $k + 1$ points such that any k of them form a proper simplex of S. For reasons that will become clear later on, I will call the first k points the *main simplex*, and the last one the *unit point*. For example, point frames for $\mathbf{T}_1$, $\mathbf{T}_2$, and $\mathbf{T}_3$ consist of, respectively, three points (pairwise unrelated), four points (no three of them collinear), and five points (no four of them on the same plane). See figure 1.

Figure 1. Point frames for $\mathbf{T}_1$, $\mathbf{T}_2$, and $\mathbf{T}_3$.

2.2. Mixed frames

A *mixed frame* for a flat set S is an arrangement $(s^0, \ldots s^\kappa, h)$ where $(s^0; \ldots s^\kappa)$ is a non-degenerate simplex of S, and h is an oriented hyperplane of S that avoids all vertices of that simplex. The hyperplane h is the *horizon* of the frame, and $(s^0; \ldots s^\kappa)$ is the *main simplex*. For example, a mixed frame for $\mathbf{T}_2$ consists of three points p, q, r forming a proper triangle, and a line l that does not pass through any of those points. See figure 2.

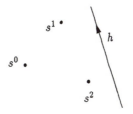

Figure 2. A mixed frame for $\mathbf{T}_2$.

In what follows we will justify these definitions, by showing that point frames and mixed frames are indeed categorical arrangements. To do that, we must examine first the conditions for two such frames to be similar.

2.3. Oriented span

Although we generally define the span of an arrangement as an *unoriented* flat set, in the case of point frames and mixed frames we can give that set an unambiguous orientation, as determined by the frame's main simplex. More precisely, if f is a point frame or mixed frame with main simplex $(s^0; \dots s^\kappa)$, then the *oriented span* of f is the flat $s^0 \vee s^1 \vee \cdots \vee s^\kappa$.

Therefore, for any *oriented* flat S, we can distinguish its *positive* point frames (whose oriented span is S) from the *negative* ones (whose oriented span is $\neg S$). Ditto for mixed frames. Note that a frame by itself is neither positive nor negative, since a positive frame for a space S is also a negative frame for the space $\neg S$.

2.4. Signature of a point frame

Let $f = (s^0, \dots s^\kappa, u)$ be a point frame. For each i in turn, consider the simplex obtained by replacing vertex s^i of the main simplex by the unit point u. Let σ_i be the orientation $(+$ or $-)$ of this simplex relative to the main one. The sequence $\sigma^0 \dots \sigma^\kappa$ is by definition the *signature* of the frame f.

In other words, the signature of a point frame is the signature of its unit point relative to its main simplex. Element σ^i of the signature tells whether u and x^i are on the same side of the hyperplane of S determined by the remaining points. In particular, the signature is $++ \cdots +$ if the unit point is inside the main simplex, and — — · · · — if it is inside the antipodal simplex. See figure 3.

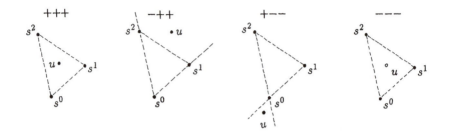

Figure 3. Point frames of $\mathbf{T}_2$ with various signatures.

2.5. Signature of a mixed frame

Let $f = (s^0, \ldots s^\kappa, h)$ be a mixed frame. The *signature* of f is the sequence $\sigma_0 \ldots \sigma_\kappa$, where $\sigma_i \in \{\pm 1\}$ is the relative position of the flats s^i and h in the oriented span of f. In other words, σ_i is such that

$$s^i \vee h = \sigma_i \circ (s^0 \vee s^1 \vee \cdots \vee s^\kappa) \quad \text{for all } i.$$

Notice how the signature of a frame f is defined solely in terms of the orientations of the elements of f, without reference to the orientation of the enclosing space. It follows that the signature is preserved by arbitrary projective maps, including those that take the oriented span of f to its opposite. As we noted before, a positive frame for a space S is also a negative frame for the space $\neg S$; but its signature is the same in both cases. In fact, for any signature σ and any space S, there are both positive and negative frames for S with signature σ. See figure 4.

Figure 4. Positive and negative frames for $\mathbf{T}_2$ with signature $+++$.

3. Standard frames

The manipulation of projective maps and frames is often simplified by the choice of "standard" reference frames for $\mathbf{T}_\nu$. For example, suppose we want to compute the matrix of the projective map that relates two arbitrary point or mixed frames a, b. One way to solve that problem is to compute the maps M_a and M_b that take some standard frame f to a and b. The desired map will then be the product $\overleftarrow{M_a} M_b$. Obviously, in order for these maps to exist, all three frames must have the same rank, type, and signature. Therefore we need at least one "standard" frame for each combination of these attributes.

3.1. Standard point frames

A standard point frame for $\mathbf{T}_\nu$ must consist of $\nu+2$ points such that any $\nu+1$ of them form a proper simplex. A natural choice is to take the main simplex $(\mathbf{e}^0; .. \mathbf{e}^\nu)$ and the *standard unit point* $\mathbf{u} = [1, 1, .. 1]$. See figures 5 and 6.

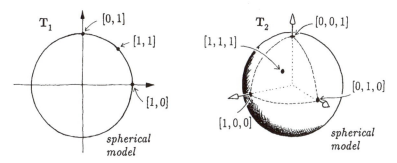

Figure 5. The standard point frames of $\mathbf{T}_1$ and $\mathbf{T}_2$ with signature $+ \cdots +$.

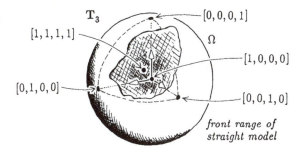

Figure 6. The standard point frame of $\mathbf{T}_3$ with signature $++++$.

In the straight model, this standard point frame consists of the front origin O, the points at infinity on each axis, and the point $(1,1,..1)$, all on the front range. This frame has signature $++\cdots+$, meaning its unit point is inside its main simplex. To obtain standard frames with other signatures, it suffices to replace u by other suitably located points. I define the *standard point frame with signature* σ, denoted by pfr_σ, as consisting of the canonical simplex $\mathbf{e}^0, ..\mathbf{e}^\nu$, plus the point $[\sigma] = [\sigma^0, ..\sigma^\nu]$ whose homogeneous coordinates are the desired signature. In the straight model, this point has Cartesian coordinates $(\sigma_1/\sigma_0, \sigma_2/\sigma_0, \ldots, \sigma_\nu/\sigma_0)$, and lies on the front or back range depending on whether σ_0 is $+1$ or -1. See figure 7.

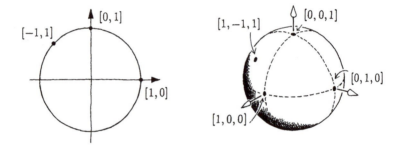

Figure 7. The standard point frames pfr_{-+} of $\mathbf{T}_1$ and pfr_{+-+} of $\mathbf{T}_2$.

3.2. Standard mixed frames

Among the mixed frames of $\mathbf{T}_\nu$ with signature $++\cdots+$, the most natural one seems to be the frame consisting of the canonical simplex together with the hyperplane $\xi = \langle 1,1..1 \rangle$. In the straight model, the hyperplane ξ is perpendicular to the vector $(1,1,..1)$ of the front range, passes through the point $(-1/\nu, -1/\nu, ..-1/\nu)$ of the negative orthant, and is oriented clockwise as seen from the front origin. See figures 8 and 9.

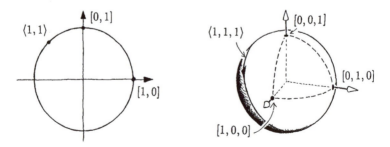

Figure 8. The standard mixed frames of $\mathbf{T}_1$ and $\mathbf{T}_2$ with all-positive signature.

As in

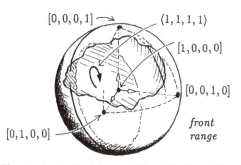

Figure 9. The standard mixed frame of $\mathbf{T}_3$ with all-positive signature.

the case of point frames, we can obtain frames of arbitrary signature by changing only the horizon hyperplane. I define the *standard mixed frame with signature* σ as consisting of the canonical simplex $(\mathbf{e}^0; ..\, \mathbf{e}^\nu)$, and the hyperplane $\langle \sigma \rangle = \langle \sigma_0, .. \sigma_\nu \rangle$ whose coefficients are the desired signature. See figure 10.

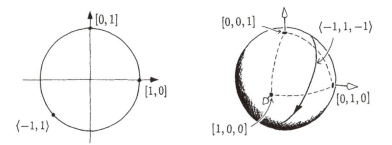

Figure 10. Some standard mixed frames of $\mathbf{T}_1$ and $\mathbf{T}_2$: (a) mfr_{-+}, (b) mfr_{-+-}.

3.3. Mapping to the standard frames

We can now prove that every point frame or mixed frame is similar to a standard one. Let's characterize first the maps that take the canonical simplex $(\mathbf{e}^0; ..\, \mathbf{e}^\nu)$ to a given simplex $s = (s^0; ..\, s^\nu)$ of $\mathbf{T}_\nu$ (positive or negative). It is not hard to see that such maps are precisely those of the form

$$
\begin{bmatrix}
\gamma_0 & & & \\
& \gamma_1 & & 0 \\
& & \ddots & \\
0 & & & \gamma_\nu
\end{bmatrix}
\begin{bmatrix}
s_0^0 & s_1^0 & \cdots & s_\nu^0 \\
s_0^1 & s_1^1 & \cdots & s_\nu^1 \\
\vdots & & & \vdots \\
s_0^\nu & s_1^\nu & \cdots & s_\nu^\nu
\end{bmatrix}
\tag{1}
$$

where the γ_i are positive but otherwise arbitrary. The theorems below show that the scale factors γ_i provide enough degrees of freedom for handling the last item (unit point or horizon) of the two frames.

Theorem 2. *Every point frame of* $\mathbf{T}_\nu$ *is similar to the standard one with the same signature.*

PROOF: Let $f = (s^0, \ldots s^\nu, u)$ be a point frame with signature $\sigma = \sigma_0, \ldots \sigma_\nu$. Let $s^i = [s^i_0, \ldots s^i_\nu]$ and $u = [u^0, \ldots u^\nu]$, and let $(\alpha_0, \ldots \alpha_\nu)$ be the solution to the linear system

$$(\alpha_0, \ldots \alpha_\nu) \begin{pmatrix} s^0_0 & \cdots & s^0_\nu \\ \vdots & & \vdots \\ s^\nu_0 & \cdots & s^\nu_\nu \end{pmatrix} = (u_0, \ldots u_\nu) \tag{2}$$

Note that since $(s^0; \ldots s^\nu)$ is a non-degenerate simplex, the matrix above has non-zero determinant. According to Cramer's rule the solution of (2) is

$$\alpha_i = \begin{vmatrix} s^0_0 & \cdots \cdots \cdots & s^0_\nu \\ \vdots & & \vdots \\ u_0 & \cdots \cdots \cdots & u_\nu \\ \vdots & & \vdots \\ s^\nu_0 & \cdots \cdots \cdots & s^\nu_\nu \end{vmatrix} \Bigg/ \begin{vmatrix} s^0_0 & \cdots \cdots \cdots & s^0_\nu \\ \vdots & & \vdots \\ s^i_0 & \cdots \cdots \cdots & s^i_\nu \\ \vdots & & \vdots \\ s^\nu_0 & \cdots \cdots \cdots & s^\nu_\nu \end{vmatrix} \tag{3}$$

Note that the signs of the numerator and denominator in (3) give the orientation of the simplices $(s^0; \ldots s^{i-1}; u; s^{i+1}; \ldots s^\nu)$ and $(s^0; \ldots s^\nu)$. It follows that $|\alpha_i| \neq 0$, and $\text{sign}(\alpha_i) = \sigma_i$.

Now consider the map M_f of the form (1) with $\gamma_i = |\alpha_i|$:

$$M = \begin{bmatrix} \begin{bmatrix} |\alpha_0| & & 0 \\ & \ddots & \\ 0 & & |\alpha_\nu| \end{bmatrix} \begin{bmatrix} s^0_0 & \cdots & s^0_\nu \\ \vdots & & \vdots \\ s^\nu_0 & \cdots & s^\nu_\nu \end{bmatrix} \end{bmatrix} \tag{4}$$

This map obviously takes the main simplex of frame pfr_σ to that of frame f. I claim that M_f also takes the unit point $[\sigma]$ of pfr_σ to the unit point u of f;

indeed,

$$[\sigma_0, .. \sigma_\nu] M_f = [\sigma_0, .. \sigma_\nu] \begin{bmatrix} |\alpha_0| & & 0 \\ & \ddots & \\ 0 & & |\alpha_\nu| \end{bmatrix} \begin{bmatrix} s_0^0 & \cdots & s_\nu^0 \\ \vdots & & \vdots \\ s_0^\nu & \cdots & s_\nu^\nu \end{bmatrix}$$

$$= [\alpha_0, .. \alpha_\nu] \begin{bmatrix} s_0^0 & \cdots & s_\nu^0 \\ \vdots & & \vdots \\ s_0^\nu & \cdots & s_\nu^\nu \end{bmatrix} = [u_0, .. u_\nu]$$

QED.

Theorem 3. *Every mixed frame of* $\mathbf{T}_\nu$ *is similar to the standard one with same signature.*

PROOF: Let $f = (s^0, .. s^\nu, h)$ be a mixed frame of $\mathbf{T}_\nu$ with signature $\sigma = \sigma_0 \cdots \sigma_\nu$. Let $s^i = [s_0^i, .. s_\nu^i]$ and $h = \langle h^0, .. h^\nu \rangle$. Define

$$\lambda = \text{sign}(s^0, .. s^\nu) \qquad \beta_i = \frac{1}{\sum_j s_j^i h^j} \tag{5}$$

Observe that the sign of β_i is the relative orientation of s^i and h in $\mathbf{T}_\nu$, and that the numerator λ is the orientation of the span of f relative to $\mathbf{T}_\nu$. Therefore, $\text{sign}(\beta_i) = \lambda \sigma_i$.

Now consider the map M_f of the form (1) with $\gamma_i = |\beta_i|$:

$$M_f = \begin{bmatrix} |\beta_0| & & 0 \\ & \ddots & \\ 0 & & |\beta_\nu| \end{bmatrix} \begin{bmatrix} s_0^0 & \cdots & s_\nu^0 \\ \vdots & & \vdots \\ s_0^\nu & \cdots & s_\nu^\nu \end{bmatrix} \tag{6}$$

I claim that M_f takes the standard mixed frame mfr_σ to f. Obviously, M_f takes the main simplex of the former to that of the latter. As for the horizon, in order to prove that $\langle \sigma \rangle M_f = h$ it is sufficient to show that, for every point $x \in \mathbf{T}_\nu$, we have $x \diamond (\langle \sigma \rangle M_f) = x \diamond h$. Since M_f is one-to-one, we can replace x in this formula by by $x M_f$. Since M_f is a map from $\mathbf{T}_\nu$ to $\lambda \circ \mathbf{T}_\nu$, we have $(x M_f) \diamond (\langle \sigma \rangle M_f) = \lambda(x \diamond \langle \sigma \rangle)$. Therefore, all we have to show is that

$$\lambda(x \diamond \langle \sigma \rangle) = (x M_f) \diamond h \quad \text{for all } x \in \mathbf{T}_\nu. \tag{7}$$

Now, on the one hand

$$\lambda(x \diamond \langle\sigma\rangle) = \lambda(\text{sign} \sum_i x_i \sigma_i) \tag{8}$$

and, on the other hand,

$$
\begin{aligned}
(x M_f) \diamond h &= \text{sign} \sum_j (x M_f)_j h^j = \\
&= \text{sign} \sum_j (\sum_i x_i |\beta_i| s^i_j) h^j \\
&= \text{sign} \sum_i x_i |\beta_i| (\sum_j s^i_j h^j) \\
&= \text{sign} \sum_i x_i |\beta_i| \frac{\lambda}{\beta_i} \\
&= \text{sign} \sum_i x_i \lambda \, \text{sign}(\beta_i) \\
&= \lambda(\text{sign} \sum_i x_i \sigma_i)
\end{aligned}
\tag{9}
$$

By comparing (8) and (9), we conclude that indeed $\langle\sigma\rangle M_f = h$. QED.

Recall that any κ-dimensional subspace of $\mathbf{T}_\nu$ can be projectively mapped to $\mathbf{T}_\kappa$, and that projective maps are closed under inversion and composition. Thanks to these results, we can extend theorems 2 and 3 to arbitrary frames and spaces:

Theorem 4. *Two point frames or two mixed frames are similar if and only if they have the same rank and same signature.*

Note that the map M_f defined in theorems 2 and 3 is insensitive to the orientation of the unit point u or the horizon h.

3.4. Uniqueness of projective maps

It is now time to prove that the objects we have been studying so far are indeed frames. We must verify that

Theorem 5. *Point frames and mixed frames are categorical.*

PROOF: Recall that an arrangement a is categorical if the only projective map on $Span(a)$ that takes a to itself is the identity. Let's first show that every standard point frame f of $\mathbf{T}_\nu$ is categorical.

So, suppose M is a map from $\mathbf{T}_\nu$ to $\mathbf{T}_\nu$ (or $\neg \mathbf{T}_\nu$) that takes the frame f to itself. In particular, M must take the canonical simplex of $\mathbf{T}_\nu$ to itself. It follows that its matrix must have positive coefficients in the main diagonal, and must be zero everywhere else:

$$M = \begin{bmatrix} \gamma_0 & & 0 \\ & \ddots & \\ 0 & & \gamma_\nu \end{bmatrix} \tag{10}$$

Let σ be the signature of the frame f. If f is a point frame, then M must take the point $[\sigma] = [\sigma_0, .. \sigma_\nu]$ to itself. According to (10), $[\sigma]M = [\gamma_0\sigma_0 .. \gamma_\nu\sigma_\nu]$. It follows that $[\sigma]M = [\sigma]$ implies all γ_i must be equal.

Similarly, if f is a mixed frame, the map M must keep the hyperplane $\langle\sigma\rangle = \langle\sigma_0, .. \sigma_\nu\rangle$ fixed. The image of $\langle\sigma\rangle$ has coefficients $\langle\sigma\bar{M}\rangle$, where $\bar{M}$ is the adjoint of the matrix of M. The adjoint of (10) is

$$\bar{M} = \begin{pmatrix} \varepsilon_0 & & 0 \\ & \ddots & \\ 0 & & \varepsilon_\nu \end{pmatrix}$$

where $\varepsilon_i = \prod_{j\neq i} \gamma_j$. Therefore, $\langle\sigma\rangle M = \langle\varepsilon_0\sigma_0, .. \varepsilon_\nu\sigma_\nu\rangle$. In order to have $\langle\sigma\rangle M = \langle\sigma\rangle$ it is necessary that all ε_i, and therefore all γ_i, be the same.

In either case, we have shown that the matrix of M must be a positive multiple of the identity, and therefore M must be the identity map of $\mathbf{T}_\nu$. We conclude that the standard point frames and mixed frames are categorical. Because of theorem 4, it follows that all point frames and mixed frames are categorical. QED.

3.5. Computational considerations

The proofs of theorems 2 and 3 give practical methods for computing the map M that takes a standard frame to a given frame f. The matrix of M consists of the coordinates of the main simplex of f, with each row scaled by an appropriate factor.

In the case of a point frame, the factors α_i are found by solving the linear system (2). (As a byproduct, the signs of the α_i give the signature of the frame f). The linear system can be solved in $O(n^3)$ operations, by factoring the matrix (s^i_j) into the Gaussian LU product (or any similar factorization). Scaling each row of L by the corresponding factor $|\alpha_i|$ gives the desired map M, already in factored form.

In the case of a mixed frame, the scale factors α_i are given by equation (5), and the matrix of M by (1), in $O(n^2)$ operations. Note that we do not need to know the orientation λ of the frame f.

From these remarks it would seem that point frames are substantially more expensive to handle than mixed frames. However, as discussed in a previous chapter, whenever one needs a projective map one usually needs also its inverse, or some factorization of the map that gives both at roughly the same cost. Since the practical algorithms for computing these things require $\Theta(n^3)$ operations, the difference in cost between the two types of frames is actually very small.

3.6. A note on one-dimensional frames

Recall that in spaces of dimension 1, points and hyperplanes are the same thing. Therefore, in those spaces point frames are indistinguishable from mixed frames. This coincidence gives rise to some ambiguity, since the signature of a one-dimensional frame f (three collinear points) depends on whether we look at it as a point frame or as a mixed frame. For example, the frame pfr_{-+} (figure 7(a)) has signature $-+$ if viewed as a point frame, and $++$ if viewed as a mixed frame.

However, this ambiguity is of no consequence: for any 1-dimensional frame f (i.e., for any sequence of three independent collinear points) there is only one frame g of $\mathbf{T}_1$ whose main simplex is $(\mathbf{e}^0, \mathbf{e}^1)$ and is similar to f. The frame g is both a standard point frame and a standard mixed frame of $\mathbf{T}_1$, and in either interpretation it has the same signature as f. As shown below in section 3.4, there is a unique projective map that takes the three points of one frame to those of the other.

In conclusion, three points on a line determine the same map from that line to $\mathbf{T}_1$, whether we use the formulas of theorem 2, or those of theorem 3.

4. Coordinates relative to a frame

In Euclidean geometry, coordinate frames are used not so much to define maps as to assign numerical coordinates to every point. Projective frames too have that function: as shown below, a projective frame for a space S assigns to each point of S a homogeneous coordinate tuple.

By definition, the *coordinates of a point p relative to a frame f* are the homogeneous coordinates of the point pN, where $N = \overleftarrow{M}_f$ is the projective map that takes f to the appropriate standard frame g of $\mathbf{T}_\nu$. The *coefficients of a hyperplane w relative to f* are defined the same way, by mapping w through M and taking the coefficients of the result.

Obviously, the coordinates of the vertices of f's main simplex, relative to f, will be $[1, 0, 0, .., 0]$, $[0, 1, 0, .., 0]$, $[0, 0, 1, .., 0]$, and so on. If f is a point frame, its unit point u will have coordinates $[\sigma_0, .. \sigma_\nu]$ relative to f. If f is a mixed frame, its horizon will have coefficients $\langle \sigma_0, .. \sigma_\nu \rangle$ relative to f. See figure 11.

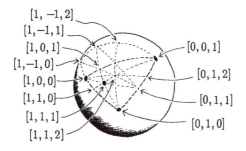

Figure 11. Coordinates relative to a given point frame.

Note that the coordinates relative to a point frame f do not depend on the orientation of the frame's unit point u. That is, if g has the same main simplex as f but unit point $\neg u$, the coordinates relative to f and relative to g are the same. Indeed, we can replace u by any of 2^n other points without affecting the map M_f or the coordinates relative to f. Those are the points whose coordinates relative to f are all ± 1.

4.1. Invariance of relative coordinates

Relative coordinates are invariant under projective maps, in the following sense. Let f be a frame, and M be the projective map that takes f to a standard frame g of $\mathbf{T}_\nu$. By definition, the coordinates of a point p relative to f are those of pM. For any map N, the map that takes frame fN to g is $\overleftarrow{N}M$. Therefore the coordinates of pN relative to the frame fN (with respect to the same standard g) are those of $(pN)\overleftarrow{N}M = pM$. That is, for any map N, the coordinates of pN relative to fN are those of p relative to f.

4.2. The center-of-mass interpretation

The coordinates relative to a point frame f have a relatively simple interpretation in terms of the straight model. Suppose that the main simplex s of f lies on the front range of the straight model. Suppose also that the frame f has signature $++\cdots+$, that is, its unit point u lies inside the simplex s. Now imagine that we place a set of "weights" $\gamma_0, .. \gamma_\nu$ at the vertices of the main simplex, in such a way that their center of mass falls on the unit point u. Then the point with relative coordinates $[x_0, .. x_\nu]$ will be the point where the center of mass will go if each weight γ_i is replaced by $\gamma_i x_i$.

In particular, when u is actually the barycenter of the simplex (whose Cartesian coordinates are the arithmetic average of the Cartesian coordinates of the vertices), the weights γ_i are all equal, and the point with relative coordinates $[x_0, \ldots x_\nu]$ will be the center of mass of weights $x_0, \ldots x_\nu$ placed at the vertices of the simplex. See figure 12. This point also has the property that its distance to the face h^i of s opposite to vertex s^i, divided by the distance from s^i to h^i, is $x_i / \sum_j x_j$.

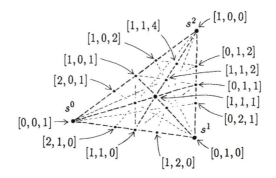

Figure 12. Barycentric coordinates.

We get this same coordinate system also when f is a mixed frame whose horizon is Ω. In either case, the resulting coordinates are called the *barycentric coordinates relative to the main simplex s*.

Chapter 13
Cross ratio

Lengths and distances play a central role in classical geometry. Indeed, Euclidean geometry can be defined as the study of those properties of figures that are preserved by rigid motions; which can be defined in turn as the transformations of space that preserve distances and lengths. The very word "geometry" reminds us that this whole area of mathematics was originally the science of measuring distances.

As we know, general projective maps do not preserve the length of segments, and not even the ratio of lengths. Still, given the importance of those concepts in Euclidean geometry, it is natural to ask whether these notions have some analogue in projective geometry. That is, is there any numerical measure of geometric objects, that we could use to express sizes and positions of points, and that is preserved by arbitrary projective maps?

The answer is yes: the so-called *cross ratio* of four points provides such a measure. In this chapter we will briefly review the properties of the classical (unoriented) cross ratio, adapt its definition to the two-sided world, and discuss its main properties in some detail.

1. Cross ratio in unoriented geometry

As we may expect, a measure that is invariant under arbitrary projective maps must be more complicated than Euclidean distance. For starters, it must have at least four arguments: one can easily show that that any function that depends on only two or three points and is invariant under arbitrary projective maps must assume only a finite set of values, and threfore is not a useful measure. (Examples of such functions are the predicates that test whether two points are antipodal and whether three points are collinear.)

In classical projective geoemtry one learns that the simplest real-valued invariant is the *cross ratio* of four collinear points, which can be defined as follows. If x, y, a, b are four distinct real numbers, then their cross ratio is the fraction

$$(x : y \mid a : b) \; = \; \frac{x - a}{b - x} \Big/ \frac{y - a}{b - y} \tag{1}$$

In general, if x, y, a, b are four distinct points on a line l, their cross ratio $(x : y \mid a : b)$ can be defined by picking an arbitrary Cartesian coordinate system on l (i.e., an origin, a direction, and a unit of length), measuring the coordinate of each point on this scale, and plugging those numbers into equation (1). See figure 1.

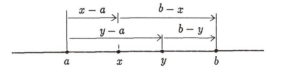

Figure 1. Cross ratio of four points.

It is easy to see that changes in the coordinate system (translations, scalings, sign reversals) do not affect the result of formula (1). It is only a little bit harder to show that formula (1) remains invariant when all four points are transformed by maps of the form $z \mapsto (\alpha z + \beta)/(\gamma z + \delta)$, which are precisely the projective maps of the real line.

The assumption that all four points are distinct can be relaxed somewhat, particularly if we accept $1/0 = \infty = -\infty$ as a valid ratio. Also, formula (1) can be continuosly extended to the cases where one of the points is at infinity. In particular, if b goes to infinity, the cross ratio reduces to the plain ratio $(x - a)/(y - a)$. It is not worth going into more details here, since we will go through a more thorough analysis for the two-sided version of cross ratio in section 2.

1.1. Interpreting the cross ratio

In order to gain a better intuition for the meaning of cross ratio, let's consider how the value of (1) varies when we keep a, y, b fixed, and move x along their common line, in the direction from a to b by way of y, possibly crossing $x = \infty$ along the way. See figure 2.

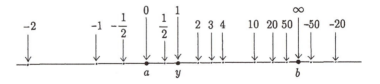

Figure 2. Values of $(x : y \mid a : b)$ as a function of x.

As x goes from a to b, the cross ratio spans the positive values from 0 (when $x = a$) to 1 (when $x = y$) to ∞ (when $x = b$). As x continues moving past b and goes back to a by the complementary route (the one not including y), the cross ratio goes through the negative values, from $-\infty$ to 0.

The point x where the cross ratio has value -1 is the *harmonic conjugate* of y with respect to a and b.

1.2. Computing the classical cross ratio

Let x be a point of the unoriented projective line $\mathbf{P}_1$ with homogeneous coordinates $[x_0, x_1]$. In the straight model, x is the point x_1/x_0 of the real line. Therefore, if we are given four distinct proper points x, y, a, b of $\mathbf{P}_1$, we can compute their classical cross ratio by the formula

$$(x : y \mid a : b) = \frac{\left(\dfrac{x_1}{x_0} - \dfrac{a_1}{a_0}\right) \Big/ \left(\dfrac{b_1}{b_0} - \dfrac{x_1}{x_0}\right)}{\left(\dfrac{y_1}{y_0} - \dfrac{a_1}{a_0}\right) \Big/ \left(\dfrac{b_1}{b_0} - \dfrac{y_1}{y_0}\right)}$$

$$= \frac{\dfrac{a_0 x_1 - a_1 x_0}{a_0 x_0} \cdot \dfrac{y_0 b_1 - y_1 b_0}{y_0 b_0}}{\dfrac{x_0 b_1 - x_1 b_0}{x_0 b_0} \cdot \dfrac{a_0 y_1 - a_1 y_0}{a_0 y_0}} = \frac{\begin{vmatrix} x_0 & x_1 \\ a_0 & a_1 \end{vmatrix} \cdot \begin{vmatrix} b_0 & b_1 \\ y_0 & y_1 \end{vmatrix}}{\begin{vmatrix} x_0 & x_1 \\ b_0 & b_1 \end{vmatrix} \cdot \begin{vmatrix} a_0 & a_1 \\ y_0 & y_1 \end{vmatrix}} \tag{2}$$

It is useful to view the cross ratio itself as a point on the projective line, whose homogeneous coordinates are

$$(x : y \mid a : b) = \left[\begin{vmatrix} x_0 & x_1 \\ b_0 & b_1 \end{vmatrix} \cdot \begin{vmatrix} a_0 & a_1 \\ y_0 & y_1 \end{vmatrix} , \begin{vmatrix} x_0 & x_1 \\ a_0 & a_1 \end{vmatrix} \cdot \begin{vmatrix} b_0 & b_1 \\ y_0 & y_1 \end{vmatrix} \right] \tag{3}$$

or, in schematic form,

$$(x : y \mid a : b) = \left[\begin{vmatrix} x \\ b \end{vmatrix} \cdot \begin{vmatrix} a \\ y \end{vmatrix} , \begin{vmatrix} x \\ a \end{vmatrix} \cdot \begin{vmatrix} b \\ y \end{vmatrix} \right] \tag{4}$$

One advantage of expressing the cross ratio in this form is that formulas (3–4) are always meaningful, even in those cases where the denominator of formula (2) is zero. (At worst, formulas (3–4) may yield the null object $\mathbf{0} = [0, 0]$ as the cross ratio.)

2. Cross ratio in the oriented framework

I will take equations (3–4) as the *definition* of cross ratio on the oriented projective line $\mathbf{T}_1$. The cross ratio itself is to be viewed as a point of $\mathbf{T}_1$, so that a cross ratio of $3/2 = [2,3]$ is distinct from a cross ratio of $(-3)/(-2) = [-2,-3] = \neg[2,3]$. This means we must be doubly careful about the order of the four points in formulas (3–4).

2.1. Interpreting the cross ratio

To appreciate the meaning of formula (4), let's consider again what happens to the cross ratio when we move x and keep a, y, b fixed, as we did in section 1.1 for the unoriented version. One difference that comes up right away is that in the oriented world y may be in four projectively distinguishable positions with respect to a and b, and the cross ratio behaves differently in each case. So, let's first consider the case where y is on the open segment ab. See figure 3.

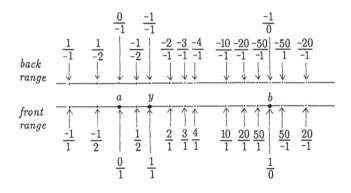

Figure 3. Values of $(x : y \mid a : b)$ as a function of x.

In that case, as x moves forward on the line $a \vee b$, the cross ratio $(x : y \mid a : b)$ moves forward on the line $\mathbf{T}_1$. Its value will be

$$\left.\begin{array}{l} \text{front positive} \\ \text{back negative} \\ \text{back positive} \\ \text{front negative} \end{array}\right\} \quad \text{if } x \text{ is between} \quad \left\{\begin{array}{l} a \text{ and } b \\ b \text{ and } \neg a \\ \neg a \text{ and } \neg b \\ \neg b \text{ and } a \end{array}\right\} \quad (5)$$

In particular, as x moves from a to b along the segment ab, the cross ratio goes from $0 = [1,0]$ (when $x = a$) to $1 = [1,1]$ (when $x = y$) to $+\infty = [0,1]$ (when $x = b$).

How does this picture change when the point y is not on the segment ab? We can answer this question by observing that both coordinates of the cross ratio are negated whenever we replace any argument by its antipode. That is,

$$
\begin{aligned}
(\neg x : y \mid a : b) &= (x : \neg y \mid a : b) \\
= (x : y \mid \neg a : b) &= (x : y \mid a : \neg b) = \neg(x : y \mid a : b)
\end{aligned}
\tag{6}
$$

Therefore, if y is on the segment from b to $\neg a$, we can understand the behavior of $(x : y \mid a : b)$ by applying the above analysis to $\neg(x : y \mid \neg a : b)$, and negating both coordinates of the resulting cross ratio. We can figure out, for example, that as x moves from $\neg a$ to b the value of $(x : y \mid a : b)$ varies from $[-1, 0]$ through $[-1, -1]$ to $[0, -1]$.

2.2. Symmetry properties

The cross ratio has a number of symmetry properties which follow directly from the defining formulas. For one thing, the cross ratio doesn't change if we swap the first pair of arguments with the second pair, or reverse the order of both pairs simultaneously. That is,

$$
(x : y \mid a : b) = (a : b \mid x : y) = (y : x \mid b : a) = (b : a \mid y : x)
\tag{7}
$$

Also, reversing only the first (or last) pair has the effect of exchanging the two coordinates of the cross ratio. If the cross-ratio is viewed as a (two-sided) real number, this operation is equivalent to taking its reciprocal. Thus, if we define $1/[r, s] = [s, r]$, we can write

$$
(y : x \mid a : b) = (x : y \mid b : a) = 1/(x : y \mid a : b)
\tag{8}
$$

Swapping the innermost (or outermost) two arguments is numerically equivalent to computing one minus the original ratio, and moving to the antipodal range. More precisely, if we define $1 - [r, s] = [r, r - s]$, then

$$
(x : a \mid y : b) = (b : y \mid a : x) = \neg\big(1 - (x : y \mid a : b)\big)
\tag{9}
$$

Combining these results, we can verify that the $4! = 24$ possible permutations split into six equivalence classes with four members each. If the original cross ratio is $\alpha = [r, s]$, the other five values that can be obtained this way are

$$
\begin{aligned}
(x : y \mid a : b) &= \alpha &&= [r, s] \\
(x : y \mid b : a) &= 1/\alpha &&= [s, r] \\
(x : a \mid y : b) &= \neg(1 - \alpha) &&= [-r, s - r] \\
(x : a \mid b : y) &= \neg 1/(1 - \alpha) &&= [s - r, -r] \\
(x : b \mid y : a) &= \neg(1 - 1/\alpha) &&= [-s, r - s] \\
(x : b \mid a : y) &= \neg 1/(1 - 1/\alpha) &&= [r - s, -s]
\end{aligned}
\tag{10}
$$

2.3. Invariance under projective maps

Lemma 1. *For any points* x, y, a, b *of* $\mathbf{T}_1$, *and any projective map* M *of* $\mathbf{T}_1$ *to* $\pm\mathbf{T}_1$, *we have*

$$(xM : yM \mid aM : zM) = (x : y \mid a : b) \tag{11}$$

PROOF: Let M be the linear map of $\mathbf{R}^2$ to itself that induces M. From definition (4), we have

$$
(xM : yM \mid aM : bM) = \left[\left| \begin{matrix} xM \\ bM \end{matrix} \right| \cdot \left| \begin{matrix} aM \\ yM \end{matrix} \right| , \; \left| \begin{matrix} xM \\ aM \end{matrix} \right| \cdot \left| \begin{matrix} bM \\ yM \end{matrix} \right| \right]
$$

$$
= \left[\left| \begin{matrix} x \\ b \end{matrix} \right| \cdot \left| \begin{matrix} a \\ y \end{matrix} \right| \cdot |M|^2 , \; \left| \begin{matrix} x \\ a \end{matrix} \right| \cdot \left| \begin{matrix} b \\ y \end{matrix} \right| \cdot |M|^2 \right] \tag{12}
$$

$$
= \left[\left| \begin{matrix} x \\ b \end{matrix} \right| \cdot \left| \begin{matrix} a \\ y \end{matrix} \right| , \; \left| \begin{matrix} x \\ a \end{matrix} \right| \cdot \left| \begin{matrix} b \\ y \end{matrix} \right| \right]
$$

$$
= (x : y \mid a : b)
$$

QED.

With this result, we can generalize the definition of cross ratio of four points x, y, a, b lying on an arbitrary two-sided line l. If φ is any isomorphism from l to $\mathbf{T}_1$, the cross ratio $(x : y \mid a : b)$ is by definition the same as $(x\varphi : y\varphi \mid a\varphi : b\varphi)$, computed according to equation (4). Lemma 1 guarantees us that the result will not depend on which isomorphism we use: if φ, η are two isomorphisms from l to $\mathbf{T}_1$, then $\overleftarrow{\varphi}\eta$ is a projective map of $\mathbf{T}_1$ to itself, which means $(x\varphi : y\varphi \mid a\varphi : b\varphi) = (x\eta : y\eta \mid a\eta : b\eta)$.

By the same argument, we can extend lemma 1 to arbitrary projective maps between arbitrary two-sided lines:

Theorem 2. *The cross ratio of four collinear points is invariant under arbitrary projective maps.*

2.4. Cross ratio as relative coordinates

Let x, y, a, b be four collinear points, with y on the segment ab. Viewed as a function of x, the cross ratio $(x : y \mid a : b)$ is a map that takes a to $[1, 0]$, y to $[1, 1]$, and b to $[0, 1]$. By inspecting formula (4), we can see that, if the other points are fixed, the coordinates of the cross ratio are homogeneous linear functions of those of x. This function is therefore the map that takes the point frame (a, b, y) to the standard point frame of $\mathbf{T}_1$ with same signature $(++)$. In other words, *the cross ratio $(x : y \mid a : b)$ gives the homogeneous coordinates of x relative to the point frame (a, b, y).*

Unfortunately, this is true only if y lies on the segment ab. In the general case, if we compute the coordinates of x relative to (a, b, y) by the formulas given in chapter 12, we get

$$
\mathrm{sign} \begin{vmatrix} a \\ b \end{vmatrix} \circ \left[\begin{vmatrix} x \\ b \end{vmatrix} \cdot \mathrm{abs} \begin{vmatrix} a \\ y \end{vmatrix}, \ \begin{vmatrix} a \\ x \end{vmatrix} \cdot \mathrm{abs} \begin{vmatrix} y \\ b \end{vmatrix} \right] \tag{13}
$$

Comparing this with formula (4), we can see that the two agree only when

$$
\mathrm{sign} \begin{vmatrix} a \\ b \end{vmatrix} = \mathrm{sign} \begin{vmatrix} a \\ y \end{vmatrix} = \mathrm{sign} \begin{vmatrix} y \\ b \end{vmatrix} \tag{14}
$$

that is, when y is in the segment ab.

The difference between formulas (4) and (13) arises because of the decision we made in chapter 12 to use the canonical simplex of $\mathbf{T}_\nu$ as the the main simplex of all standard frames. As a consequence, the formulas that give the coordinates of a point relative to a frame $f = (s^0, \dots s^\nu, u)$ treat the unit point u of f differently from the other points. Specifically, replacing u by $\neg u$ in f has no effect in the map M_f, whereas replacing one of the s^i by $\neg s^i$ modifies the map so as to send $\mathbf{e}^i$ to $\neg s^i$ instead of s^i.

We could have avoided this discrepancy by taking equation (13) instead of (4) as the definition of cross ratio. However, the value computed according to formula (13) does not deserve to be called a cross ratio, since its symmetry properties are nowhere as nice as those of the classical version.

Alternatively, we could have defined the standard frames in such a way that the formulas for relative coordinates were equally sensitive to the orientation of all points of the frame, and reduced to formula (4) in the one-dimensional case. The symmetrical cross ratio would then come out as a special case of relative coordinates. The disadvantage of this approach is that it would make it harder to predict the signs of relative coordinates.

Given these trade-offs, I have chosen not to resolve the discrepancy, and leave relative coordinates and cross ratio as distinct notions. It is quite possible that further experience with these concepts will recommend a different choice.

Chapter 14
Convexity

Convexity theory provides an important example of the advantages of the two-sided approach. In affine geometry, a set of points is said to be convex if it contains every segment whose endpoints lie in the set. This notion has no clean counterpart in classical projective geometry, essentially because one cannot define unambiguously *the* segment connecting two given points. By contrast, in two-sided geometry the segment joining two points is well defined and unique, as long as they are not antipodal. Moreover, this segment can be defined solely in terms of join, and so is a purely projective concept. Therefore, in two-sided geometry there is a notion of convexity that is preserved by projective maps, and yet retains most of the properties of affine convexity. Thus, in two-sided geometry we can use the tools of projective geometry to the development of theorems and algorithms involving convexity. The result is a theory of convex sets that is cleaner and richer than the affine version.

1. Convexity in classical projective space

Attempts to extend the notion of convexity to *unoriented* projective space have followed two major approaches.

The first approach is to assign special meaning to some fixed line, say the line at infinity Ω. A convex set is then defined as one that avoids Ω, and that contains every line segment whose ends lie in the set and which does not cross Ω. The problem with this solution is that is doesn't give us anything new: by excluding Ω we are simply working on the affine plane. In particular, convexity will be preserved only by those projective maps that take Ω to itself, i.e. the affine maps. Therefore, this solution negates most of the advantages of projective geometry.

A second approach (used by Sylvester) is to say that a set X is convex if there is *some* line l disjoint from X such that every line segment with ends in X that does not cross l is contained in X. What differs from the previous approach is that the line l is allowed to depend on X. As a result, this brand of convexity is preserved by projective maps; unfortunately, it lacks other properties we usually

131

associate with convexity, such as closure under intersection. As figure 1 shows, the intersection of two convex figures may fail to be convex, or even connected.

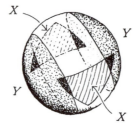

Figure 1. Intersecting convex sets of $\mathbf{P}_2$.

2. Convexity in oriented projective spaces

In *oriented* projective geometry we can use the affine definition of convexity almost without change: a set of points is convex if it contains every segment whose endpoints lie in the set. However, before we use this definition we have to clarify the meaning of "segment," especially in the case of antipodal endpoints.

2.1. Open and closed segments

Recall that in chapter 4 we defined the open segment pq, for two distinct and non-antipodal points of $\mathbf{T}_\nu$, as the set of all points x such that $(p; x)$ and $(x; q)$ are proper simplices equivalent to $(p; q)$; or, equivalently, that $p \lor r = r \lor q = p \lor q \neq \mathbf{0}$. For the purposes of the this chapter, it is convenient to define also the open segments pp and $p(\neg p)$ as being the empty set.

Let's also define the *closed segment* pq as being the open segment pq together with the points p and q. In particular, if $p = q$ the closed segment is a single point and if $p = \neg q$ it is just the set $\{p, q\}$.

2.2. Convex sets of points

Now that we know what a segment is, let's go back to our proposed definition of convexity, which says that a set of points is convex if it contains all open segments whose endpoints are in the set. What should we do about antipodal points? We get two different notions of convexity, depending on whether we allow or exclude such pairs:

Definition 1. A subset X of a two-sided space is *convex* if X contains every segment whose endpoints are in X.

Definition 2. A subset X of a two-sided space is *strictly convex* if it is convex and contains no antipodal pairs of points.

Note that for subsets contained in the front range of $\mathbf{T}_\nu$, the definitions 1 and 2 are equivalent, and describe precisely those subsets of the front range that are convex in the sense of classical affine geometry.

The two-sided space of zero rank $\mathbf{T}_{-1}$ has no points, and therefore its only convex set is the empty set. The space $\mathbf{T}_0$ of rank one is only a bit more interesting: the set of points is $\{\Upsilon_0, \neg\Upsilon_0\}$, all open segments are empty, and all sets of points are convex.

In the two-sided line $\mathbf{T}_1$ a convex set must be one of the following: the empty set, a single point, a pair or antipodal points, a segment with two independent endpoints (including none, one, or both of them), a half-line (including none, one, or both of its antipodal endpoints), or the whole $\mathbf{T}_1$.

Examples of strictly convex subsets of $\mathbf{T}_2$ are the empty set, a single point, the interior of a proper simplex, an open half-space, an open wedge (the intersection of two half-spaces), and half of a straight line with at most one of its endpoints. Also, any subset of the front or back range of the straight model that is convex in the affine sense is also a strictly convex set by definition 2. However, these sets are not the only possibilities: as figure 2 shows, a strictly convex set of $\mathbf{T}_\nu$ may extend across both ranges (and may seem seem rather non-convex to a naive observer.)

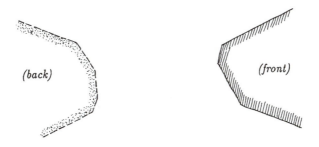

Figure 2. A strictly convex subset of $\mathbf{T}_2$, in the straight model.

Some examples of convex sets that are not strictly convex are: a straight line, a closed half-space, a pair of antipodal points, the whole two-sided space, and one half of a straight line with both its endpoints.

3. Properties of convex sets

3.1. Invariance under projective maps

Note that since segments can be defined in terms of join, they are defined on arbitrary two-sided spaces, not just on the standard space $\mathbf{T}_\nu$. Therefore, the same observation applies to the notions of convexity and strict convexity. Furthermore, since isomorphisms by definition commute with the join operation, they must take segments to segments. We can immediately conclude that

Theorem 1. *Convexity and strict convexity are preserved by isomorphisms between projective spaces, in particular by projective maps of $\mathbf{T}_\nu$ to itself.*

Therefore, any projective properties of convex subsets that hold in the standard space $\mathbf{T}_\nu$ hold also on any two-sided space of the same dimension.

3.2. Intersection properties

Theorem 2. *The convex subsets of a two-sided space are closed under arbitrary intersections.*

PROOF: Consider an arbitrary, non-empty family of convex subsets of some space S, and let X be their intersection. Let p, q be any two points of X. Clearly, p and q must be in every member of the family. Since each member is convex, it must contain the segment pq. Therefore, that segments must also be contained in X. QED.

The following is an obvious corollary:

Theorem 3. *Strictly convex sets are closed under arbitrary intersections.*

The next two theorems show that convex sets can be recognized by their one-dimensional sections, just as in affine geometry:

Theorem 4. *A subset of a two-sided space S is convex if and only if its intersection with every line of S is convex.*

PROOF: The "only if" part is a corollary of theorem 2, since a line of $\mathbf{T}_\nu$ is a convex set of points. As for the converse, suppose the set X is not convex. Then there are two points p, q in X such that the open segment pq is not contained in X. Obviously, p and q must be distinct and not antipodal. Let l be the line $p \vee q$.

The intersection $X \cap l$ contains the points p and q, but not the segment pq, which is not contained in X. Therefore, $X \cap l$ is not convex.
QED.

Theorem 5. *A subset X of a two-sided space S is strictly convex if and only if it its intersection with every line of S is strictly convex.*

Theorem 5 is an immediate corollary of theorem 4. Note that a strictly convex subset of a line must be one of these: the empty set; a point; a proper line segment with zero, one, or two of its endpoints; or a half-line with at most one of its endpoints.

3.3. Topological properties

Recall that a two-sided space S of rank n has the topology of the sphere $\mathbf{S}_\nu$. This topology defines the open and closed subsets of S, and also defines the interior, exterior, boundary, and closure of any subset X of S, relative to S.

Theorem 6. *In a space S of positive rank, A convex subset has empty interior relative to S if and only if it is contained in some hyperplane of S.*

PROOF: The "if" part is trivial, since the hyperplanes of S have no interior points. For the "only if" part, let X be a convex subset of S that is not contained in any hyperplane. The set X must therefore include n points that form a proper ν-dimensional simplex s. If $n \geq 2$, the points inside s (those with signature $++\cdots+$) are contained in X, by convexity, and are easily shown to be interior points of X. If $n = 1$, the space S has the discrete topology (every single point is both an open and a closed set), so the only vertex of s is an interior point of X. In any case, we conclude that a convex set not contained in any hyperplane must have a non-empty interior.
QED.

Theorem 7. *In a two-sided space S, a convex subset X with non-empty interior (relative to S) is contained in the closure of its interior.*

PROOF: The theorem is trivially true when $\text{rank}(S) \leq 1$, so let's suppose $\text{rank}(S) = n \geq 2$. Let p be any point of X. If X has non-empty interior, it is not contained in any hyperplane of S. Therefore, there must be $n - 1$ other points of X that together with p form a proper $(n-1)$-dimensional simplex s. By convexity, the interior I_s of this simplex is contained in the interior I_X of X. Point p is in the closure of I_s, and therefore in the closure of I_X. Since this holds for all $p \in X$, the theorem is proved.
QED.

Theorem 8. *Any proper subset of a space S that is open and convex is strictly convex.*

PROOF: Let X be such a set, and suppose it contained two antipodal points $p, \neg p$. Since X is open, p must lie in some open neighborhood $N \subseteq X$. Let l be any line through p. The line l includes $\neg p$, p, and points of $N \cap l$ (an open subset of l) that approach p on both sides. Since X is convex, it follows that $l \cap X$ must be the whole l. Since this holds for all lines through p, X must be the whole space, contradicting the hypothesis. We conclude that X has no antipodal points, and therefore is strictly convex.
QED.

Informally, theorems 7 and 8 say that convex sets either are entirely flat, or don't have any flat parts. In the second case, the set must consist of an open strictly convex set, plus some subset of its boundary.

3.4. Analytic characterization of convexity

Analytically, a point r is on the closed segment pq if and only if its *homogeneous* coordinates are non-negative linear combinations of those of p and q. If $p = [x]$, $q = [y]$, and $r = [z]$, this condition means $z_i = \alpha x_i + \beta y_i$, for all i and for some $\alpha, \beta \geq 0$. In fact, we can conclude that r is on the closed segment pq if and only if $r = p$, $r = q$, or there are $x, y \in \mathbf{R}^n$ such that $p = [x]$, $q = [y]$, and $r = [x+y]$. It follows that

Theorem 9. *A subset X of $\mathbf{T}_\nu$ is convex if and only if it contains the point $[x+y]$, for any independent points $[x], [y] \in X$.*

In the spherical model, we can see that a closed segment of $\mathbf{R}^n$ that does not contain the origin $\vec{0} = (0, .. 0)$ is mapped by central projection onto a segment of the two-sided plane. This gives us two additional characterizations of convex sets:

Theorem 10. *A subset X of $\mathbf{T}_\nu$ is convex if and only if it is the central projection on $\mathbf{S}_\nu$ of $Y \setminus \vec{0}$, where Y is a convex subset of $\mathbf{R}^n$.*

PROOF: If Y is a convex subset of $\mathbf{R}^n$, the convexity of its projection X follows from the definition and from the observations in the preceding paragraph.

Conversely, let X be a convex subset of $\mathbf{T}_\nu$. Let's consider X as a subset of the unit sphere of $\mathbf{R}^n$, and let Y be the union of all closed segments connecting the origin of $\mathbf{R}^n$ to a point of X. See figure 3. It is easy to check that Y is convex; and, obviously, the central projection of $Y \setminus \vec{0}$ on $\mathbf{S}_2$ is X.

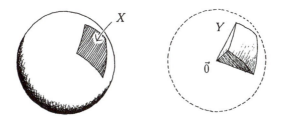

Figure 3. Convexity in $\mathbf{T}_\nu$ and in $\mathbf{R}^n$.

QED.

Theorem 11. *A subset X of $\mathbf{T}_\nu$ is strictly convex if and only if it is the central projection of a convex subset of $\mathbf{R}^n \setminus \vec{0}$.*

3.5. Convexity and projective maps

Among the many trivial corollaries of theorems 10 and 11, we have:

Theorem 12. *If X is a convex set of points and F is any projective function defined on X, then $F(X \setminus Null(F))$ is convex.*

PROOF: Let X be a convex subset of $\mathbf{T}_\nu$, and F a projective function of $\mathbf{T}_\nu$ to $\mathbf{T}_\kappa$. Following theorem 10, let Y be a convex subset of $\mathbf{R}^n$ such that X is the central projection of $Y \setminus \vec{0}$. Let F be a function of $\mathbf{R}^n$ to $\mathbf{R}^k$ that generates the projective function F. Then $F(X \setminus Null(F))$ is the central projection of $F((Y \setminus \vec{0}) \setminus Null(F)) = F(Y \setminus Null(F))$. By the definition of null space, $F(Y \setminus Null(F)) = F(Y) \setminus \vec{0}$. Since linear functions preserve convexity in $\mathbf{R}^n$, $F(Y)$ is convex. By theorem 10, it follows that X is convex.

QED.

Theorem 13. *If X is a strictly convex set, F is a projective function defined on X, and X is disjoint from $Null(F)$, then $F(X)$ is strictly convex.*

PROOF: Following theorem 11, let $X \subseteq \mathbf{T}_\nu$ be the central projection of $Y \subseteq \mathbf{R}^n \setminus \vec{0}$. Let $F = [\![F]\!]$ where F is a linear function from $\mathbf{R}^n$ into $\mathbf{R}^k$. Then $F(X)$ is the central projection of $F(Y)$. From $X \cap Null(F) = \emptyset$ we have $Y \cap Null(F) = \emptyset$, and therefore $F(Y) \subseteq \mathbf{R}^k \setminus \vec{0}$. By theorem 11, $F(X)$ is strictly convex.

QED.

4. The half-space property

The set of all points in a two-sided space is convex, but not strictly convex. How big can a strictly convex set be? Roughly speaking, not bigger than half of the containing space. In this section we will make this statement more precise.

4.1. Supporting half-spaces

Lemma 14. *Every open or closed strictly convex subset of a space S is contained in the positive side of some hyperplane of S.*

PROOF: Let X be an open strictly convex subset of S. We will prove the theorem by induction on the rank n of S. We can assume X to be non-empty, for otherwise the theorem is trivial; hence the rank n is at least 1.

If $n = 1$, S has only two points, the universe $\varUpsilon$ and its opposite. In that case we must have $X = \{\varUpsilon\}$ or $X = \{\neg\varUpsilon\}$. Then the hyperplane $\varLambda$ or $\neg\varLambda$, respectively, will leave X on its positive side.

Now suppose $n \geq 2$. If X is strictly convex, it must be a proper subset of S, and disjoint from its own antipodal image $\neg X$. Its set-theoretic complement $S\backslash X$ is a proper nonempty subset of S. Since S is connected, X and $S \setminus X$ cannot be both open or both closed. On the other hand, the map $x \mapsto \neg x$ is continuous and one-to-one, so X and $\neg X$ are both open or both closed. We conclude that $\neg X$ is a proper subset of $S \setminus X$. That is, there is some point p that is neither in X nor in $\neg X$. Obviously, the same is true of $\neg p$.

Now let π be a right complement of p in S, that is, a hyperplane such that $p \vee \pi = S$. Let F be the projection of S from p onto π:

$$F(x) = (p \vee x) \wedge \pi \tag{1}$$

See figure 4.

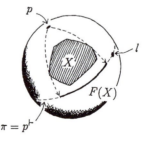

Figure 4.

The null space of F is the set $\{p, \neg p\}$, and its range is the whole hyperplane π. Since $X \subseteq S \setminus \mathit{Null}(F)$, by theorems 1 and 2 of chapter 11 the set $F(X)$ is open if X is open, and closed if X is closed. According to theorem 13, $F(X)$ is a strictly convex subset of π. By induction, there must be a hyperplane l of π that leaves $F(X)$ on its positive side. That is, for all $x \in X$ we must have $F(x) \vee l = \pi$. Therefore $p \vee F(x) \vee l = p \vee \pi = S$. Since $p \vee F(x) = p \vee x$, we have $p \vee x \vee l = S$, which implies $x \vee \neg(p \vee l) = S$. Since this holds for all $x \in X$, we conclude that X lies in the positive side of the hyperplane $\neg(p \vee l)$.

QED.

4.2. Perfect half-spaces

Note that theorem 14 applies only to strictly convex sets that are either open or closed. For general strictly convex sets the theorem is false: a counterexample is the subset of $\mathbf{T}_1$ consisting of one half circle plus one of its endpoints. To fix this problem, let's define a *perfect half* of a space S recursively as being the empty set, if $\mathrm{rank}(S) = 0$, or an open half-space of S, plus a perfect half of its bounding hyperplane, if $\mathrm{rank}(S) > 0$.

For example, a *perfect half-line* consist of an open half-line plus exactly one of its endpoints. A *perfect half-plane* of $\mathbf{T}_2$ consists of an open hemisphere of $\mathbf{S}_2$, an open half-line on its boundary, and one endpoint of that half-line. See figure 5.

Figure 5. A perfect half-plane of $\mathbf{T}_2$.

The perfect halves of a space S are all projectively equivalent. However, if $\mathrm{rank}(S) \geq 1$, the perfect halves can be divided in two classes according to their "handedness," that is, according to whether the map relating them is positive or negative. For example, we can classify the perfect half-lines of $\mathbf{T}_1$ according to whether they are open at the "forward" or at the "backward" end, as seen from any interior point. Note that the antipodal image of a perfect half H of S is another perfect half of S (with same or opposite handedness, depending on whether S has even or odd rank), and is exactly the same as the set-theoretic complement $S \setminus H$.

It is not hard to prove recursively that the intersection of a perfect half-space of S with any flat a of S is a perfect half of a. It is easy to see that perfect half-lines are strictly convex. Because of theorem 5, it follows that any perfect half-space is strictly convex. Since a perfect half-space includes one member of every antipodal pair, it cannot be augmented without ceasing to be convex.

In other words, perfect half-spaces are *maximal strictly convex sets*. The converse is also true: every maximal strictly convex subset of a space S is a perfect half of S. This result is a trivial consequence of the following theorem:

Theorem 15. *Every strictly convex subset of a space S is contained in a perfect half of S.*

PROOF: The proof is by induction on the rank of S. If S has rank zero, it has no points, and the theorem is vacuously true. So, suppose S has rank $n \geq 1$, and let X be a strictly convex subset of S.

If X has no interior points, then by theorem 6 it is contained in some hyperplane f. If X has some interior points, then its interior is a strictly convex open set. By theorem 14, the interior of X is contained in the positive side of some hyperplane f. By theorem 7, X is contained in the closure of the interior of X, and hence it is contained in the closure of f's positive side.

In either case we conclude that X is contained in the union of some hyperplane f and the positive side of f. Now $X \cap f$ is a strictly convex subset of f; by induction, $X \cap f$ is contained in some perfect half H of f. Therefore, X is contained in the perfect half of S consisting of H and the positive side of f.
QED.

In fact, we can prove a stronger result:

Theorem 16. *Any strictly convex subset X of a space S is the intersection of all perfect half-spaces of S that contain X.*

PROOF: By theorem 15, there is at least one perfect half-space containing X, so the intersection Y of all such half-spaces is well-defined. Obviously, X is contained in Y. To show that $Y \subseteq X$, we have to prove that for every point $p \notin X$, there is a perfect half-space that contains X but not p.

Consider such a point p. If $\neg p \in X$, then any perfect half-space that includes X automatically excludes p, and we are done. If $\neg p \notin X$, then by theorem 13 the projection of X onto any hyperplane h complementary to p is a strictly convex subset of h. By theorem 15, it is contained in some perfect half H of h. Let Y be the set of all points of S that project onto H, plus the point $\neg p$. It is easy to check that Y contains X but not p, and is a perfect half of S.
QED.

From theorem 17 we can derive similar results for open and closed convex sets, in terms of open and closed half-spaces:

Theorem 17. *Any open convex subset X of a space S is the intersection of S and all open half-spaces of S that contain X.*

PROOF: If X is empty, or is the whole of S, the theorem is trivially true. Otherwise, by theorem 8 the set X is strictly convex; by theorem 16, it is the intersection of all perfect half-spaces that contain X. Since X is open, it is contained in the interior of every one of those perfect half-spaces; but those interiors are exactly the open half-spaces that contain X.
QED.

Theorem 18. *Any closed convex subset X of a space S is the intersection of S and all closed half-spaces of S that contain X.*

PROOF: We can prove this result by induction on the rank of S. If X is empty, or is the whole of S, (in particular, if $\text{rank}(S) = 0$), the theorem is trivially true. If X has empty interior, theorem 6 says that X is contained in some hyperplane of S, and the result follows by induction. Otherwise, by theorem 17 the interior X' of X is the intersection of all open half-spaces that contain X'; therefore, the closure X'' of X' is the intersection of all closed half-spaces that contain X''. Since $X'' = X$, by theorem 7, the result follows.
QED.

5. The convex hull

The *convex hull* of an arbitrary set X of points of a space S is the intersection $Hull(X)$ of all convex subsets of S that contain X. Since S itself is one such subset, the intersection is always well-defined. See figure 6.

Figure 6. A subset of $\mathbf{T}_2$ (hatched) and its convex hull.

5.1. Convex union

The *convex union* $X \uplus Y$ of two arbitrary sets of points X, Y is by definition $Hull(X \cup Y)$, the convex hull of their set-theoretical union. In other words,
$X \uplus Y$ is the smallest convex set that contains both X and Y. This operation is
obviously commutative, and it is easy to check that it is also associative; that is,

$$X \uplus (Y \uplus Z) = (X \uplus Y) \uplus Z = Hull(X \cup Y \cup Z) \tag{2}$$

Since every set X is the union of its single-element subsets, it follows that the hull
of X is the convex join of those subsets:

$$Hull(X) = \biguplus_{x \in X} \{x\} \tag{3}$$

5.2. Convex polytopes

A strictly convex set that is the convex hull of a finite set of points is called
a *convex polytope*. This general term includes the two-dimensional *convex polygons*
and the three dimensional *convex polyhedra* as special cases.

It can be shown that the boundary of a convex κ-dimensional polytope
can be partitioned into a finite number of open convex polytopes of dimension 0
through $\kappa - 1$, with pairwise disjoint relative interiors and spanning pairwise distinct
flats; these are the *faces* of the polytope. The *facial lattice* of the polytope is the
finite combinatorial structure consisting of those faces and the set inclusion relations
between them.

There is a vast literature on convex polytopes of and their facial lattices,
and a proper treatment of the subject would take at least another book. Readers
interested in this subject should start from the books by Grünbaum [11], Yaglom
and Boltyanskii [7], and Coxeter [20], for a mathematical exposition; or the one by
Edelsbrunner [9], for a computationally oriented view.

Note that a convex polytope, being strictly convex and closed, must be contained in an open half-space of $\mathbf{T}_\nu$. If we identify that open half-space with the
classical affine space $\mathbf{A}_\nu$, we conclude that every convex polytope of $\mathbf{T}_\nu$ is projectively equivalent to a convex polytope of classical affine geometry. Therefore, all the
combinatorial results about facial lattices that can be found in the literature remain
valid in two-sided geometry.

6. Convexity and duality

So far we have discussed convexity only for sets of points, but of course we can dualize all our definitions and theorems to work with sets of sets of hyperplanes instead. In this and the following sections we explore the consequences of this duality.

6.1. Dual segments

If a, b are two independent hyperplanes of $\mathbf{T}_\nu$, then the *dual open segment ab* determined by them is the set of all hyperplanes h such that $a \wedge h = h \wedge b = a \wedge b$. For example, in $\mathbf{T}_2$ this dual segment is a fan of oriented lines that pass through the point $a \wedge b$ and whose directions span the shorter arc between those of a and b. In $\mathbf{T}_3$, the dual open segment is the "book" consisting of those planes that contain the common line $a \wedge b$ and whose normals, at any fixed point on that line, span the shorter angle between the normals of a and b. See figure 7.

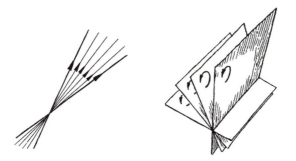

Figure 7. Dual segments in $\mathbf{T}_2$ and $\mathbf{T}_3$.

As in the primal case, when $a = b$ or $a = \neg b$ it is convenient to define the dual open segment ab as being the empty set.

6.2. Convex sets of hyperplanes

A set H of hyperplanes is *convex* if it contains the dual open segment ab for any hyperplanes a, b of H.

Obviously, convex sets of points and convex sets of hyperplanes are dual notions. If η is any duomorphism from a two-sided space S to another space T, and a, b are any two points of S, the image under η of the open segment ab is the dual open segment of T determined by the hyperplanes a^η and b^η. If C is a convex set of points of S, it then follows trivially that the image $C^\eta = \{\, c^\eta \; : \; c \in C \,\}$ is a convex set of hyperplanes of T. Dually, η also maps a convex set of hyperplanes of S to a convex set of points of T.

All projective properties of convex sets of points can therefore be dualized into properties of convex sets of hyperplanes. In particular, the *dual convex hull* of a set H of hyperplanes of S is the smallest convex set $Hull(H)$ of hyperplanes that contains H.

The left half of figure 8 shows a convex set C of points of $\mathbf{T}_2$, and the right half shows some of the elements of $C^{\dashv}$, a convex set of lines that is the image of C under the standard duomorphism $\dashv$ of $\mathbf{T}_2$.

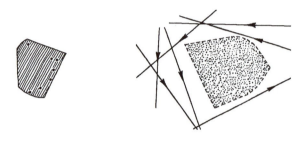

Figure 8.

A detailed analysis of this example would show that $C^{\dashv}$ is exactly the set of those lines of $\mathbf{T}_2$ that pass to the left of a convex region C^*, shown shaded in the right-hand figure. The relation between this region and the original set C is far from trivial; for instance, its corners correspond to the straight edges of C, and vice-versa. As we shall see in the following sections, this region plays an important role in the theory of convex sets.

6.3. Support and kernel

Let X be an arbitrary set of points of $\mathbf{T}_\nu$. By definition, the (*closed*) *left support* and the (*closed*) *right support* of X are the sets

$$X^{\dashv} = \{\, y \ : \ \forall x \in X \quad y \diamond x \geq 0 \,\} \tag{4}$$

$$X^{\triangleright} = \{\, y \ : \ \forall x \in X \quad x \diamond y \geq 0 \,\} \tag{5}$$

In words, a hyperplane y is in the left support of a point set X if and only if it either contains or is a left complement of every point x of X.

Dually, if X is an arbitrary set of hyperplanes of $\mathbf{T}_\nu$, equations (4) and (5) define the (*closed*) *left kernel* and the (*closed*) *right kernel* of X. In words, a point y is in the left kernel of X if and only if it either lies on or is a left complement of every hyperplane of X.

What is the relation between $X^{\triangleleft}$ and $X^{\triangleright}$? Recall that $a \diamond b = (-1)^{rs}(b \diamond a)$, where $r = \mathrm{rank}(a)$ and $s = \mathrm{rank}(b)$. When one flat is a point and the other is a hyperplane of $\mathbf{T}_\nu$, we have $rs = n - 1 = \nu$, so

$$X^{\triangleright} = \neg^\nu X^{\triangleleft} \tag{6}$$

That is, $X^{\triangleright} = X^{\triangleleft}$ in spaces of even dimension, and $X^{\triangleright} = \neg X^{\triangleleft}$ in spaces of odd dimension.

Note that the the functions $\triangleright$ and $\triangleright$ do not act elementwise (as $\dashv$ and $\vdash$ do), but depend on the given set as a whole. For instance, the set $(X \cup Y)^{\triangleright}$ is not necessarily equal to $(X^{\triangleright} \cup Y^{\triangleright})$. In fact, it follows from the definition that

$$(X \cup Y)^{\triangleright} = X^{\triangleright} \cap Y^{\triangleright}$$

In particular, if X is not empty,

$$X^{\triangleright} = \bigcap_{x \in X} \{x\}^{\triangleright} \tag{7}$$

If x is a hyperplane, the set $\{x\}^{\triangleright}$ is one of the two closed half-spaces delimited by x, namely the the set of all points y such that $x \diamond y \geq 0$. Therefore, equation (7) says that the right kernel $X^{\triangleright}$ of an arbitrary set of hyperplanes X is the intersection of a family of closed convex sets of points. Dually, the right support of an arbitrary set of points is the intersection of a family of closed convex sets of hyperplanes. Since the intersection of any number of closed convex sets is a closed convex set, we conclude that

Theorem 19. *For any set of points or hyperplanes X, the sets $X^{\triangleleft}$ and $X^{\triangleright}$ are closed and convex.*

Needless to say, the functions $\triangleleft$ and $\triangleright$ commute with negation:

$$\neg(X^{\triangleleft}) = (\neg X)^{\triangleleft} \qquad \neg(X^{\triangleright}) = (\neg X)^{\triangleright} \tag{8}$$

Moreover, they commute with the polar complements $\dashv$ and $\vdash$:

Theorem 20. *For any set X of points or hyperplanes in $\mathbf{T}_\nu$,*

$$X^{\vdash\triangleleft} = X^{\triangleleft\vdash} = X^{\dashv\triangleright} = X^{\triangleright\dashv}$$
$$X^{\vdash\triangleright} = X^{\triangleright\vdash} = X^{\dashv\triangleleft} = X^{\triangleleft\dashv}$$
$$X^{\vdash\triangleleft} = \neg^\nu X^{\vdash\triangleright}$$

PROOF: The equality $X^{\vdash\triangleleft} = X^{\triangleleft\vdash}$ follows from the definitions, via a double renaming of variables and the observation that $a \diamond b = a^{\dashv} \diamond b^{\dashv}$:

$$
\begin{aligned}
X^{\vdash\triangleleft} &= \{ y \ : \ \forall x \in X^{\vdash} \quad y \diamond x \geq 0 \} \\
&= \{ y \ : \ \forall x \in X^{\vdash} \quad y^{\dashv} \diamond x^{\dashv} \geq 0 \} \\
&= \{ y \ : \ \forall x \in X \quad y^{\dashv} \diamond x \geq 0 \} \\
&= \{ y^{\vdash} \ : \ \forall x \in X \quad y \diamond x \geq 0 \} \\
&= X^{\triangleleft\vdash}
\end{aligned}
$$

A dual argument shows that $X^{\dashv\triangleright} = X^{\triangleright\dashv}$. The remaining equalities then follow from the formulas $X^{\triangleleft} = \neg^{\nu} X^{\triangleright}$ and $X^{\vdash} = \neg^{\nu} X^{\dashv}$.
QED.

6.4. Convex hull as kernel of support

Given an arbitrary set X of points or hyperplanes $\mathbf{T}_{\nu}$, consider the set $X^{\triangleleft\triangleright}$ (which, according to equation(6), is the same as $X^{\triangleright\triangleleft}$). Theorem 19 tells us that this is a closed convex set, whose elements have the same rank (1 or $n-1$) as those of X.

Since every element x of X satisfies $y \diamond x \geq 0$, for every element y of $X^{\triangleleft}$, we conclude that the original set X is a subset of $X^{\triangleleft\triangleright}$. In fact, when X is a set of points, $X^{\triangleleft\triangleright}$ is the kernel of the support of X, that is, the intersection of all closed half-spaces of $\mathbf{T}_{\nu}$ that contain X. By theorem 18 (and its dual), we conclude that $X^{\triangleleft\triangleright} = X$ whenever X is a closed convex set.

From equation (7) it follows that, for any sets X and Y,

$$ X \subseteq Y \Rightarrow X^{\triangleleft} \supseteq Y^{\triangleleft} $$

and, therefore,

$$ X \subseteq Y \Rightarrow X^{\triangleleft\triangleright} \subseteq Y^{\triangleleft\triangleright} $$

In particular, if Y is any closed convex set that contains X, we have $X^{\triangleleft\triangleright} \subseteq Y^{\triangleleft\triangleright} = Y$.

We conclude that $X^{\triangleleft\triangleright}$ is the intersection of all closed convex sets that contain X, which is also the intersection of all convex sets that contain the closure of X. We have thus proved

Theorem 21. *For any set X of points or hyperplanes of $\mathbf{T}_{\nu}$, $X^{\triangleleft\triangleright} = X^{\triangleright\triangleleft} = Hull(Clos(X))$.*

6.5. Convex duality

We now have two distinct ways of transforming a convex sets of points into a convex set of hyperplanes, and vice versa. One way is to map the set pointwise through a duomorphism, such as $\dashv$ or $\vdash$; The other way is to use the closed support/kernel functions $\lhd$ and $\rhd$.

While the correspondence defined by a duomorphism is one-to-one and can be applied to arbitrary sets of flats, the one defined by $\lhd$ and $\rhd$ is one-to-one only when restricted to closed convex sets. To see why, recall from section 5 that $X^{\lhd\rhd} = X^{\rhd\lhd} = X$ when X is a closed convex set; that is, $\rhd$ and $\lhd$ are inverses of each other for such sets, and are therefore one-to-one. On the other hand, if X is not closed and convex, we will have $X^{\lhd\rhd} = Y^{\lhd\rhd}$ for any other set Y that contains X and is contained in $Hull(Clos(X))$, which shows that those functions are not one-to-one in general.

Both kinds of "duality" have the drawback that they map sets of points into sets of hyperplanes, which are somewhat hard to visualize. However, by composing these two dualities, we get a new mapping that takes point sets to point sets, and is still in some sense a "duality" between such sets.

For any closed convex set C of points of $\mathbf{T}_\nu$, consider the set $C^* = C^{\vdash\lhd}$, the pointwise right complement of the closed left support of C. This set is the intersection of all the closed half-planes $\{c\}^\vdash$ where c ranges over the points of C. I call this set the *convex dual* of set C. In figure 8, the shaded region at the right is the convex dual C^* of the hatched region C on the left.

According to theorem 20, the set C^* can be written in many ways:

$$C^{\vdash\lhd} = C^{\lhd\vdash} = C^{\dashv\rhd} = C^{\rhd\dashv} = C^*$$
$$C^{\vdash\rhd} = C^{\rhd\vdash} = C^{\dashv\lhd} = C^{\lhd\dashv} = \neg^{n-1}C^*$$

As the following theorem shows, the convex duality function $*$ is a one-to-one mapping from closed convex sets to closed convex sets, and is in fact its own inverse:

Theorem 22. *For any closed convex set X of $\mathbf{T}_\nu$, we have $(C^*)^* = C$.*

PROOF: We need only apply the definition, noting that $\vdash\lhd = \rhd\dashv$ (by theorem 20) and that $C^{\lhd\rhd} = C$ (because C is closed and convex):

$$(C^*)^* = (C^{\lhd\vdash})^{\lhd\vdash} = ((C^{\lhd})^{\vdash\lhd})^\vdash = ((C^{\lhd})^{\rhd\dashv})^\vdash = (C^{\lhd\rhd})^{\dashv\vdash} = C$$

QED.

Tables(9), (10) and(11) below list some pairs of closed convex sets of points that are convex duals of each other, in spaces of various dimensions:

T_1	
empty set	whole two-sided line
single point	closed half two-sided line
two antipodal points	two antipodal points
non-trivial segment	non-trivial segment

(9)

T_2	
empty set	whole two-sided plane
single point	closed half-plane
two antipodal points	a line
non-trivial segment	wedge (angle) between two lines
closed half-line	closed half-line
triangle	triangle

(10)

T_3	
empty set	whole two-sided space
single point	closed half-space
two antipodal points	a plane
non-trivial segment	wedge (dihedron) between two planes
closed half-line	closed half-plane
closed flat figure	closed cone
closed tetrahedron	closed tetrahedron

(11)

6.6. Properties of convex duality

Recall that the kernel/support operation ◁ reverses the sense of set inclusion, that is, $X \subseteq Y$ implies $Y^{\triangleleft} \subseteq X^{\triangleleft}$. Since the convex duality $*$ is the composition of ◁ and the polar complement ⊢, and the latter maps sets elementwise, it is also the case that

$$X \subseteq Y \Rightarrow Y^* \subseteq X^* \tag{12}$$

Note that the convex duality $*$ does not commute with general projective maps and duomorphisms, but only with those that preserve the standard polarity relation $\perp$ of $\mathbf{T}_\nu$. Note also that $*$, being a correspondence between closed convex sets rather then oriented flats, does not establish a duality between $\wedge$ and $\vee$. However, it does establish a duality between two other important operations on such sets, namely intersection and convex join:

Theorem 23. *If X, Y are two closed convex sets of $\mathbf{T}_\nu$, then*

$$(X \cap Y)^* = X^* \uplus Y^*$$
$$(X \uplus Y)^* = X^* \cap Y^*$$

PROOF: For the first equation, note that $X \cap Y$ is a subset of both X and Y; therefore, equation (12) tells us that $X^* \subseteq (X \cap Y)^*$ and $Y^* \subseteq (X \cap Y)^*$, that is, $(X \cap Y)^*$ contains $X^* \cup Y^*$. In fact, $X \cap Y$ is the *largest* closed convex set that is contained in both X and Y; since $*$ is one-to-one for closed convex sets, we conclude that $(X \cap Y)^*$ must be the *smallest* closed convex set that contains both X^* and Y^*, which is precisely $X^* \uplus Y^*$.

The second equation follows from the first and the identity $X^{**} = X$:

$$(X \uplus Y)^* = ((X^*)^* \uplus (Y^*)^*)^*$$
$$= ((X^* \cap Y^*)^*)^*$$
$$= X^* \cap Y^*$$

QED.

Recall that the convex hull of a set X can be written as the convex join of its singleton subsets. If the hull is a closed set (in particular, if the set X is finite) then we can apply theorem 23 and conclude that

$$Hull(X) = \left(\bigcap_x \in X\{x\}^* \right)^* \tag{13}$$

That is, the convex hull of a finite set of points is the convex dual of the intersection of the closed half-spaces that are the convex duals to those points.

6.7. Applications of convex duality

The convex duality ∗ is extremely important in the theory of convex poly-topes, where it is called "combinatorial duality" (when applied to facial lattices), "geometric duality," or simply "duality." The facial lattices of a polytope P and its dual P^* have the same incidence structure, except that the inclusion relations are reversed: for each face of P with rank k, the corresponding face of P^* has rank $n - k$.

In particular, the facial lattice of a three-dimensional polytope can be viewed as a planar graph, and then ∗ coincides with the graph-theoretical duality that preserves edges but exchanges vertices with faces; for instance, the convex dual of a cube is a octahedron, and that of a regular dodecahedron is an icosahedron. For more details, see the references cited in section 5.

The convex duality ∗ is also quite valuable in practice, since, like projective duality, it lets us use a single algorithm to solve two different problems. One example is given by equation (13), which lets us use the same code for computing the convex hull of a set of points and intersecting a collection of half-spaces.

Another example is computing the Voronoi diagram and the Delaunay tri-angulation of a set of points in $\mathbf{R}^\nu$. As noted by K. Q. Brown [5], if we map the given points by inverse stereographic projection onto the unit sphere of $\mathbf{T}_{\nu+1}$, and we compute the convex hull H of the resulting points, then the projection of H back onto $\mathbf{T}_\nu$ gives the desired Delaunay triangulation, while the projection of the dual H^* gives the desired Voronoi diagram.

As noted before, a major advantage of oriented projective geometry is it ability to handle both convexity and duality in a consistent framework. Although it is possible to define the convex duality ∗ in classical affine geometry (extended with the points at infinity), its domain must be restricted to those closed convex sets that include a fixed point O (the convex dual of the whole affine plane). Therefore, results such as $X \uplus Y = (X^* \cap Y^*)^*$, which are always valid in the two-sided framework, cannot be used in affine geometry if the sets involved have no common point.

Chapter 15
Affine geometry

We can use oriented projective geometry to emulate all constructions and algorithms of Euclidean geometry, by restricting our attention to the front range of the straight model of $\mathbf{T}_\nu$, An Euclidean flat is simulated by the front part of a flat of $\mathbf{T}_\nu$; Euclidean and affine transformations are provided by the projective maps that take the front range to itself; and so on. In this and the next two chapters we will discuss the details of this emulation.

When working on problems from affine and Euclidean geometry, it may seem a mathematical overkill and a waste of resources to use n homogeneous coordinates to represent a point, when $n - 1$ Cartesian coordinates would be enough. But there are several good reasons for doing so. One reason is simplicity: with homogeneous coordinates we can better handle some degenerate situations, such as computing the meet of parallel lines. This freedom to leave the confines of the Euclidean plane in intermediate computations often simplifies the programs enormously, even if the input data and the final results have to be represented in Cartesian coordinates. Another reason is standardization: life is much simpler if all geometric software in a computer system uses the same data format. A third possible reason is efficiency and accuracy: with homogeneous coordinates we can often avoid doing any divisions (which in some computers are more expensive than multiplications), except at the very end when converting the results back to Cartesian coordinates.

1. The Cartesian connection

Recall that the analytic and straight models of $\mathbf{T}_\nu$ are related by central projection

$$[x_0,\, x_1,\dots x_\nu] \quad \mapsto \quad \left(\frac{x_1}{x_0}, \frac{x_2}{x_0}, \dots, \frac{x_\nu}{x_0}\right) \tag{1}$$

By definition, the projected point lies in either the front or the back range, depending on whether x_0 is positive or negative. If x_0 is zero, the projection is the point at infinity in the direction $(x_1,\, x_2,\dots x_\nu)$. This map is the standard way of converting from homogeneous to Cartesian coordinates. The inverse map is

$$(x_1,\, x_2,\dots x_\nu) \quad \mapsto \quad [\pm 1,\, x_1,\, x_2,\dots x_\nu] \tag{2}$$

151

where the weight coordinate is $+1$ for points on the front range, and -1 for points on the back.

In principle, formulas (1) and (2) are all we need to emulate Cartesian and Euclidean geometry in the two-sided framework. However, as the next few sections will make clear, there are a few subtle points in the handling of signs which require some careful thought.

1.1. The midpoint of a segment

For example, let $p = [u, a, b]$ and $q = [v, c, d]$ be points on the front range of $\mathbf{T}_2$. The midpoint of the segment pq has Cartesian coordinates

$$\frac{1}{2}\left(\frac{a}{u} + \frac{c}{v}, \ \frac{b}{u} + \frac{d}{v}\right) = \frac{1}{2uv}(va + uc, \ vb + ud)$$

Converting this point back to homogeneous coordinates, we finally get

$$\text{midpoint}([u, a, b], [v, c, d]) \ = \ [\, 2uv, \ va + uc, \ vb + ud\,] \tag{3}$$

1.2. Natural or absolute?

Formula (3) works fine as long as p and q lie on the front range, but may fail in other cases. Since the coordinates of the midpoint in formula (3) are homogeneous linear functions of the coordinates of p (or q), replacing p or q by its antipode has the same effect on the result. Therefore, if p and q are both on the back range, the result of (3) lies on the *front* range, and indeed is the midpoint of $\neg p$ and $\neg q$. Worse still, when p and q are on opposite ranges, formula (3) gives a point on the back range that is the midpoint of either p and $\neg q$, or of $\neg p$ and q.

If we don't like this behavior, we can replace formula (3) by

$$\text{midpoint}(\,[u, a, b], [v, c, d]\,) \ = \ \big[\,|v|\,u + |u|\,v, \ |v|\,a + |u|\,c, \ |v|\,b + |u|\,d\,\big] \tag{4}$$

This formula gives always a point on the segment pq, provided p and q are not both at infinity, and $p \neq \neg q$. When p and q are on the same range, the result is the midpoint of pq. When p and q are on opposite ranges, the result is the point where that segment crosses the line at infinity.

Thee dilemma presented by formulas (3) and (4) is similar to the one we encoutered while defining cross-ratio in chapter 13. This problem occurs over and over again when deriving formulas for two-sided geometry. A straightforward way of deriving such a formula is to compute the Cartesian coordinates of the operands, apply the appropriate formula from Cartesian geometry, convert the result back to homogeneous coordinates, and eliminate any divisions by a suitable rescaling of the latter. However, the "natural" formula we obtain through this route often behaves as desired only for points on the front range.

In many cases, we can fix this natural formula — usually, by inserting some absolute-value operations in the right places — so that it gives satisfactory results for points on the back range, too. However, the resulting "absolute" formula is usually not linear in the coordinates of the operands, and therefore its result changes in a complicated way when the operands' orientations are reversed.

This dilemma has no universal solution: the proper choice between natural and absolute formulas usually depends on the application. However, for low-level general-purpose geometric libraries, the best formulas are usually the natural ones: for this application, ensuring simple, symmetric, consistent, and mathematically elegant semantics is more important than preserving "tradition" or attempting to match the programmer's intuitive expectations. The simpler the building blocks are, the easier it is to put them together.

2. Two-sided affine spaces

The concepts of affine geometry, such as *direction, parallelism, affine map, affine ratio, midpoint,* and *barycenter,* can all be redefined in $\mathbf{T}_\nu$ in terms of meet and join, if we let the hyperplane at infinity Ω_ν of $\mathbf{T}_\nu$ play a special role.

Recall that Ω_ν was defined in chapter 7 as the flat generated by the last ν points $(\mathbf{e}^1; ..\mathbf{e}^\nu)$ of the standard simplex. Note that the point $\mathbf{e}^0$ lies on the positive side of Ω_ν. In the straight model of $\mathbf{T}_\nu$, Ω_ν is the hyperplane at infinity oriented so that its positive side is the front range of $\mathbf{T}_\nu$. In particular, Ω_2 is the line at infinity of $\mathbf{T}_2$, oriented counterclockwise as seen from the origin; and Ω_3 is the plane at infinity of $\mathbf{T}_3$, oriented so that its circular arrow turns clockwise as seen from the origin.

I will call the flats of $\mathbf{T}_\nu$ contained in Ω_ν *improper,* and all the others *proper.* Observe that, by this definition, the universe $\mathbf{T}_\nu$ is a proper flat, and the vacuum Λ is an improper one.

2.1. Directions

If a is a proper flat, then $a \wedge \Omega_\nu$ is always defined, and is an improper flat with rank one less than the rank of a. I will call this flat the *direction of a,* and denote it by dir a. Note that $\mathrm{dir}(\neg a) = \neg(\mathrm{dir}\,a)$.

For example, the direction of a proper line is the point of Ω_ν where the line exits the front half of $\mathbf{T}_\nu$. The direction of a proper plane is one of the two lines at infinity that lie on that plane, namely the one which turns around the front part of the plane in a way that agrees with the orientation of a. See figure 1.

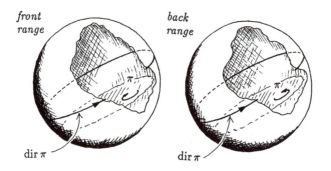

Figure 1. Direction of a plane of $\mathbf{T}_3$.

The hyperplane Ω_ν cuts every proper flat a in two open subsets, one in each range of $\mathbf{T}_\nu$. A point p will be in the front part of a if and only if $p \vee \operatorname{dir} a = a$. In particular, a point p is in the front range of $\mathbf{T}_\nu$ if and only if $p \diamond \Omega_\nu = +1$.

As for the trivial cases, the direction of a point is Λ or $\neg \Lambda$ depending on whether the point lies on the front or back range of $\mathbf{T}_\nu$; and that of Υ_ν is Ω_ν itself.

2.2. Parallelism

Two proper flats a, b are *parallel* if $\operatorname{dir} a \subseteq \operatorname{dir} b$, or $\operatorname{dir} b \subseteq \operatorname{dir} a$ (as flat sets). I will write this condition as $a \parallel b$. Note that every proper flat is parallel to itself and to its opposite, and to any of its proper sub- or super flats, including the universe Υ_ν. Also, every proper point is parallel to every proper flat.

The predicate $a \parallel b$ is symmetric, but in general not transitive: if a line l is parallel to a plane π, and π is parallel to another line m, it doesn't follow that $l \parallel m$. However, from $a \parallel b$ and $b \parallel c$ we *can* deduce $a \parallel c$, provided that $\operatorname{rank}(a) \geq \operatorname{rank}(b) \geq \operatorname{rank}(c)$ or $\operatorname{rank}(a) \leq \operatorname{rank}(b) \leq \operatorname{rank}(c)$.

If two flats a, b have the same rank k, their directions have same rank $k-1$. In this situation we can distinguish the case where a and b are *co-parallel* ($\operatorname{dir} a = \operatorname{dir} b$) from the case where they are *contra-parallel* ($\operatorname{dir} a = \neg \operatorname{dir} b$). I will denote these conditions by $a \Uparrow b$ and $a \Updownarrow b$, respectively.

If f is a proper flat, and p is any point on the front range, then $p \vee \operatorname{dir} f = p \vee (f \wedge \Omega_\nu)$ is the (unique) flat with same rank as f that is co-parallel to f and passes through p. If p lies on the back range, this formula gives a flat that is contra-parallel to f and passes through p. This formula is the two-sided analogue of an important construction of classical geometry, drawing a parallel to a given line twhough a given point.

2.3. General two-sided affine spaces

Observe how the definitions of direction and parallelism given above can be expressed only in terms of meet, join, and the special hyperplane Ω_ν. We can threfore extend those definitions to any projective space S, by using any fixed hyperplane h of S in lieu of Ω_ν. This remark motivates the following definitions:

Definition 1. A *two-sided affine space* is a pair $A = (S, h)$ where S is an oriented projective space and h one of its hyperplanes, the *horizon* of A.

Definition 2. The *canonical two-sided affine space* $\mathbf{A}_\nu$ consists of the space $\mathbf{T}_\nu$, with Ω_ν as the horizon.

From now on, by *affine space* I will usually mean a two-sided one. When necessary, I will say the *one-sided affine space* to refer to the classical one.

All notions that can be defined for $\mathbf{T}_\nu$ in terms of meet, join, and Ω_ν can automatically be translated to an arbitrary affine space $A = (S, h)$, in the obvious way. For example, if f is a proper flat of A (a sub-flat of S that is not contained in h), the *direction of f in A* is $\mathrm{dir}_A(f) = f \wedge_S h$. Two flats a, b in S are *co-parallel in A*, denoted by $a \mathbin{\Uparrow_A} b$, iff $a \wedge_S h = b \wedge_S h$. The *front range* of A is the positive side of h relative to S, that is, the set of points p such that $p \vee h = S$. Their antipodals constitute the *back range* of A.

2.4. Subspaces of an affine space

If $A = (S, h)$ is an affine space, and f is a proper flat of A, then $F = (f, f \wedge_S h)$ is also an affine space: the *affine subspace of A induced by f*. The front range of F is the part of f contained in the front range of A. It is not hard to check that the functions dir_B, $\Uparrow_B$ and $\|_B$ in the subspace B of A induced by f are merely dir_A, $\Uparrow_A$ and $\|_A$ restricted to the subflats of f.

The affine subspaces of $\mathbf{A}_\nu = (\mathbf{T}_\nu, \Omega_\nu)$ have the form $(f, \mathrm{dir}(f))$, where f is a flat not contained in Ω_ν. A two-dimensional affine subspace of $\mathbf{A}_3$, for example, consists of a proper plane π and one of its two lines at infinity, whose orientation must agree with the circular arrow on π as seen from the front range of $\mathbf{T}_\nu$. See figure 1. For the rest of this chapter, we will generally identify a flat f of $\mathbf{T}_\nu$ with the affine subspace of $\mathbf{A}_\nu$ induced by f.

For each affine space $A = (S, h)$ there are three other spaces which differ from A only in the orientation of their parts, namely $(\neg S, h)$, $(S, \neg h)$, and $(\neg S, \neg h)$. Since $\neg S \wedge h = \neg h$, only the last of these three, which I denote by $\neg A$, is a subspace of A (and vice-versa) The other two alternatives, $(\neg S, h)$ and $(S, \neg h)$ are subspaces of each other. Their front and back ranges are switched with respect to A, and their dir functions always return a result opposite to that of dir_A.

2.5. Affine maps

Informally, an affine map is a geometric transformation that preserves parallelism. In the two-sided framework, I will define an *affine map* between two affine spaces $A = (S, h)$ and $B = (T, k)$ as a projective map M from S to T that takes h to k. This last condition implies that M takes the front range of A to the front range of B, and that M commutes with the functions dir, $\parallel$, $\Uparrow$, and $\Updownarrow$, relative to A and B. That is,

$$\mathop{\mathrm{dir}}_{B}(fM) = (\mathop{\mathrm{dir}}_{A}(f))M$$

$$(fM) \Uparrow_B (gM) \Leftrightarrow f \Uparrow_A g$$

$$(fM) \Updownarrow_B (gM) \Leftrightarrow f \Updownarrow_A g$$

$$(fM) \parallel_B (gM) \Leftrightarrow f \parallel_A g$$

Let's consider in particular the affine maps of $\mathbf{A}_\nu$ to itself. Recall that a projective map M of $\mathbf{T}_\nu$ to $\mathbf{T}_\nu$ is characterized by an $n \times m$ matrix M with positive determinant. To be an affine map, M must take the improper points $\mathbf{e}^1, \ldots \mathbf{e}^\nu$ of $\mathbf{A}_\nu$ to improper points; therefore, it must be of the form

$$M = \begin{bmatrix} m_0^0 & m_1^0 & m_2^0 & \cdots & m_\nu^0 \\ 0 & m_1^1 & m_2^1 & \cdots & m_\nu^1 \\ 0 & m_1^2 & m_2^2 & \cdots & m_\nu^2 \\ \vdots & \vdots & \vdots & & \vdots \\ 0 & m_1^\nu & m_2^\nu & \cdots & m_\nu^\nu \end{bmatrix} \tag{5}$$

Moreover, since M maps Ω_ν to itself, it must map the front origin $O = \mathbf{e}^0$ to a point on the front range. We conclude that

$$m_0^0 > 0 \tag{6}$$

and, given that det M is positive,

$$\begin{vmatrix} m_1^1 & \cdots & m_\nu^1 \\ \vdots & & \vdots \\ m_1^\nu & \cdots & m_\nu^\nu \end{vmatrix} > 0 \tag{7}$$

Conversely, it is easy to check that any map of the form (5) satisfying (6–7) is an affine map of $\mathbf{A}_\nu$ to itself. For example,

$$
\begin{bmatrix}
1 & 3 & 2 \\
0 & 2 & 0 \\
0 & 1 & 3
\end{bmatrix}
\tag{8}
$$

is an affine map of $\mathbf{A}_2$ to $\mathbf{A}_2$ whose effect is depicted in figure 2.

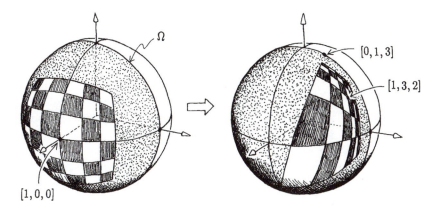

Figure 2. An affine mapping of the plane.

In general, a projective map with the form of (5) takes $\mathbf{A}_\nu$ to one of the four affine spaces $(\mathbf{T}_\nu, \Omega_\nu)$, $(\neg\mathbf{T}_\nu, \Omega_\nu)$, $(\mathbf{T}_\nu, \neg\Omega_\nu)$, or $(\neg\mathbf{T}_\nu, \neg\Omega_\nu)$. If $m_0^0 > 0$, then the front part of $\mathbf{A}_\nu$ is mapped into itself; this means the map takes $\mathbf{A}_\nu$ to either $\mathbf{A}_\nu$ or $\neg\mathbf{A}_\nu$, depending on the sign of the cofactor (7). Conversely, if $m_0^0 < 0$, the map exchanges the front and back ranges of $\mathbf{A}_\nu$; this means it takes $\mathbf{A}_\nu$ either to the affine space $(\neg\mathbf{T}_\nu, \Omega_\nu)$ or to $(\mathbf{T}_\nu, \neg\Omega_\nu)$, depending on the sign of the cofactor (7).

For example, consider the map

$$
N =
\begin{bmatrix}
-1 & 0 & \cdots & 0 \\
0 & -1 & & 0 \\
\vdots & & \ddots & \vdots \\
0 & 0 & \cdots & -1
\end{bmatrix}
\tag{9}
$$

This map takes every point of $\mathbf{T}_\nu$ to its antipode, and therefore every flat f of rank r to the flat $\neg^r f$. In particular, N takes $\mathbf{T}_\nu$ to $\neg^n \mathbf{T}_\nu$, and Ω_ν to $\neg^{n-1}\Omega_\nu$; that is, it takes the affine space $\mathbf{A}_\nu$ to either $(\mathbf{T}_\nu, \neg\Omega_\nu)$ or $(\neg\mathbf{T}_\nu, \Omega_\nu)$, depending on whether ν is odd or even.

2.6. Affine frames

An *affine frame* for an affine space $A = (S, h)$ is simply a mixed frame for S whose horizon is h. Since h is already implicit in the space, an affine frame for A is just a proper simplex spanning S with no vertices on h.

Let $s = (s^0, \ldots s^\kappa)$ and $t = (t^0, \ldots t^\kappa)$ be affine frames for two affine spaces $A = (S, h)$ and $B = (T, g)$. Is there some affine map from A to B that takes s to t? As we know from the properties of mixed frames, there exists a projective map from S to T that takes s to t and h to g if and only if the two frames have the same signature: that is, if $s^i \diamond_S h = t^i \diamond_T g$ for all i. When these conditions are satisfied, the map exists and is unique.

How do we compute this map? In the most general case, where S and T are κ-dimensional flats of $\mathbf{T}_\mu$ and $\mathbf{T}_\nu$, and the horizons h and g are arbitrary, we have to compute the projective map relating two mixed frames, a problem we already solved in chapter 12. As we discussed in that chapter, it is convenient to break down the problem into two steps, by computing maps M_s and M_t from some "standard" frame f of $\mathbf{T}_\kappa$ to the frames s and t, respectively. The desired map then will be the composition $\overleftarrow{M}_s M_t$. In particular, we may let f be the standard mixed frame with the proper signature, and compute M_s and M_t by formula (6) from chapter 12.

2.7. Affine maps between subspaces of $\mathbf{A}_\nu$

The formulas of chapter 12 become a little simpler if A and B are affine subspaces of $\mathbf{A}_\nu$ and $\mathbf{A}_\mu$, respectively; that is, if $h = S \wedge \Omega_\nu$ and $g = T \wedge \Omega_\mu$. In that case, the signature σ of the frame s tells whether each point is on the front or back range of $\mathbf{T}_\nu$; that is, $\sigma_i = \mathrm{sign}(s^i_0)$. The map that takes the standard mixed frame mfr_σ of $\mathbf{T}_\kappa$ to s is

$$
M_s = \left[\begin{array}{ccc} |1/s^0_0| & & 0 \\ & \ddots & \\ 0 & & |1/s^\kappa_0| \end{array} \right] \left[\begin{array}{ccc} s^0_0 & \cdots & s^0_\nu \\ \vdots & & \vdots \\ s^\kappa_0 & \cdots & s^\kappa_\nu \end{array} \right] \tag{10}
$$

The homogeneous coordinates determined by this map on the space A are the *barycentric coordinates* relative to the simplex s, as defined in chapter 12.

2.8. Standard affine frames

A minor inconvenience of using mfr_σ as the intermediate frame is that it is not an affine frame of $\mathbf{A}_\kappa$ (recall that the horizon of mfr_κ is $\langle\sigma\rangle$, rather than Ω_ν). Therefore, the map M_s of (10) does not take the affine space $\mathbf{A}_\kappa$ to A, but rather the space $(\mathbf{T}_\kappa, \langle\sigma\rangle)$. If we use formula (10) only as a tool for computing maps between two affine subspaces A and B, the fact that M_s is not affine is irrelevant, since the composition $\overleftarrow{M_s} M_t$ will be.

However, suppose that for some reason we need a map like M_s all by itself, and we want it to be affine. Then, instead of the standard mixed frame of $\mathbf{T}_\kappa$, we can use the *standard affine frame* of $\mathbf{A}_\kappa$, the simplex consisting of the origin

$$\mathbf{u}^0 = [1, 0, 0, 0, 0, ..] = O$$

and the κ *axial units*

$$\begin{aligned}
\mathbf{u}^1 &= [1, 1, 0, 0, 0, ..] \\
\mathbf{u}^2 &= [1, 0, 1, 0, 0, ..] \\
\mathbf{u}^3 &= [1, 0, 0, 1, 0, ..] \\
&\vdots
\end{aligned} \tag{11}$$

The points $\mathbf{u}^1, .. \mathbf{u}^\kappa$ are the points at unit distance from the origin on each coordinate axis of the front range. See figure 3.

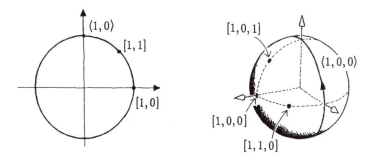

Figure 3. The standard affine frames of $\mathbf{T}_1$ and $\mathbf{T}_2$.

Actually, this is only the first member of a family of 2^n standard affine frames. In general, the standard affine frame with signature σ, denoted by afr_σ, consists of the points $\sigma_0 \mathbf{u}^0, \sigma_1 \mathbf{u}^1, .. \sigma_\nu \mathbf{u}^\nu$.

2.9. The map determined by an affine frame

Suppose we are given an affine frame s with signature σ for an affine space $A = (S, h)$, and we need to compute the map M_s from $\mathbf{A}_\kappa$ to A that takes the standard affine frame afr_σ to s. We can derive the formula for M_s by relating both afr_σ and s to the standard mixed frame of $\mathbf{T}_\kappa$. That is, we compose the projective map taking afr_σ to mfr_σ with the map taking mfr_σ to $(s^0, \ldots s^\kappa, h)$. The former is a simple map that depends only on ν and σ, and the latter is given by formula (6) of chapter 12.

In particular, if A is an affine subspace of $\mathbf{A}_\nu$, the map M_s turns out to be

$$M_s = \begin{bmatrix} 1/s_0^0 & 0 & 0 & \cdots & 0 \\ -1/s_0^0 & 1/s_0^1 & 0 & \cdots & 0 \\ -1/s_0^0 & 0 & 1/s_0^2 & & \vdots \\ \vdots & \vdots & & \ddots & \vdots \\ -1/s_0^0 & 0 & \cdots & & 1/s_0^\kappa \end{bmatrix} \begin{bmatrix} s_0^0 & s_1^0 & \cdots \cdots \cdots & s_\nu^0 \\ s_0^1 & s_1^1 & \cdots \cdots \cdots & s_\nu^1 \\ \vdots & \vdots & & \vdots \\ \vdots & \vdots & & \vdots \\ s_0^\kappa & s_1^\kappa & \cdots \cdots \cdots & s_\nu^\kappa \end{bmatrix}$$

$$= \begin{bmatrix} 1 & \dfrac{s_1^0}{s_0^0} & \dfrac{s_2^0}{s_0^0} & \cdots & \dfrac{s_\nu^0}{s_0^0} \\ 0 & \dfrac{s_1^1}{s_0^1} - \dfrac{s_1^0}{s_0^0} & \dfrac{s_2^1}{s_0^1} - \dfrac{s_2^0}{s_0^0} & \cdots & \dfrac{s_\nu^1}{s_0^1} - \dfrac{s_\nu^0}{s_0^0} \\ \vdots & \vdots & \vdots & & \vdots \\ 0 & \dfrac{s_1^\kappa}{s_0^\kappa} - \dfrac{s_1^0}{s_0^0} & \dfrac{s_2^\kappa}{s_0^\kappa} - \dfrac{s_2^0}{s_0^0} & \cdots & \dfrac{s_\nu^\kappa}{s_0^\kappa} - \dfrac{s_\nu^0}{s_0^0} \end{bmatrix} \qquad (12)$$

The relative coordinates determined by this map are the *affine coordinates relative to the frame s*. In this coordinate system, point s^i gets coordinates $\sigma_i \circ \mathbf{u}^i$. The weight coordinate of a point will be positive or negative depending on whether the point is on the front or back part of S.

2.10. Affine interpolation

An important application of affine coordinates is the problem of *affine interpolation*: given two points a, b on the front range of $\mathbf{T}_\nu$, divide the segment ab in two parts whose lengths (in the straight model) are in a given ratio $\lambda : 1 - \lambda$. In other words, find the point c on the segment ab that is λ of the way from a to b. See figure 4.

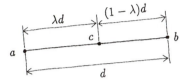

Figure 4. Affine interpolation.

This is a generalization of the midpoint problem we discussed briefly in section 1.2. One way to approach this problem is to find an affine map M, from $\mathbf{A}_1$ onto the line $a \vee b$, that takes the standard affine basis of $\mathbf{A}_1$ to the pair a, b; that is, $[1, 0] \mapsto a$ and $[1, 1] \mapsto b$. Then the point λ the way from a to b is the image of point $[1, \lambda]$ by the map M.

According to (12), the map M satisfying these conditions is

$$
M_{(a;b)} = \begin{bmatrix} 1 & \dfrac{a_1}{a_0} & \dfrac{a_2}{a_0} & \cdots & \dfrac{a_\nu}{a_0} \\[2ex] 0 & \dfrac{b_1}{b_0} - \dfrac{a_1}{a_0} & \dfrac{b_2}{b_0} - \dfrac{a_2}{a_0} & \cdots & \dfrac{b_\nu}{b_0} - \dfrac{a_\nu}{a_0} \end{bmatrix} \tag{13}
$$

The point c will then be

$$
c = [1, \lambda] M
$$

$$
= \left[1, \; \dfrac{a_1}{a_0} + \lambda \left(\dfrac{b_1}{b_0} - \dfrac{a_1}{a_0} \right), \; .. \; \dfrac{a_\nu}{a_0} + \lambda \left(\dfrac{b_\nu}{b_0} - \dfrac{a_\nu}{a_0} \right) \right] \tag{14}
$$

$$
= [\, a_0 b_0, \; \lambda a_0 b_1 + (1 - \lambda) b_0 a_1, \; .. \; \lambda a_0 b_\nu + (1 - \lambda) b_0 a_\nu \,] \tag{15}
$$

which is also what we would get by interpolating the Cartesian coordinates in the normal way. Observe that in formula (15) the purpose of multiplying the coordinates of a by b_0, and those of b by a_0, is to normalize both points to have the same weight $a_0 b_0$. After this normalization the homogeneous coordinates from 1 to ν are the same as the Cartesian ones, expressed in a common scale. That being the case, we can obtain the desired point c by interpolating the *homogeneous* coordinates in the given ratio $\lambda : 1 - \lambda$.

2.11. Absolute interpolation

Note that formula (15) is multilinear in the coordinates of the given points. It is therefore a generalization of the "natural" midpoint formula (3), and, like it, gives the intuitively correct result only if a and b lie on the front range. (Note that this assumption was necessary to ensure that the frame a, b had positive signature, so that $[1, 0]$ and $[1, 1]$ got mapped to a and b by $M_{(a;b)}$. It was also used in the passage from equation (14) to (15), which is valid only if the product $a_0 b_0$ is positive.) When a and/or b lie on the back range, neither (14) nor (15) give the result one would expect. The first formula always puts c on the front range, while the second puts it on the front if a and b lie on the same range, and on the back if they lie on opposite ranges. So, for instance, the "midpoint" of a and b, computed by either formula, may even fail to be on the segment ab. This is hardly what one would expect of an interpolation algorithm.

Is there a formula that agrees more closely with the intuitive idea of interpolation? If a and b are both on the back range, we might just take the antipodal of formula (14) or (15). That corresponds to replacing the map $M_{(a;b)}$ by $\neg M_{(a;b)}$, the map from $\mathbf{T}_1$ to the affine line $a \vee b$ with horizon $\neg \operatorname{dir}(a \vee b)$ instead of $\operatorname{dir}(a \vee b)$. However, when a and b lie on opposite ranges, we must give up any hope of using an affine map for interpolation, since some point λ of $\mathbf{T}_1$ in the finite range $0 < \lambda < 1$ must be mapped to a point at infinity, and no affine map can do that.

This problem is one we already encountered in chapters 12 and 13, and stems from the way the standard affine frame afr_σ was defined. Recall that the horizon of afr_σ is always Ω_ν, while the vertices of the main simplex may be $\mathbf{u}^i$ or $\neg\mathbf{u}^i$ as necessary to give the desired signature. If we want to have the interval from 0 to 1 mapped to the segment ab (or in general, the interior of the simplex $(u^0; .. u^\kappa)$ mapped to the interior of the given simplex s), we must do the opposite: we must use a standard frame whose main simplex stays fixed at $(u^0; .. u^\kappa)$, and whose horizon varies according to the desired signature.

It is easy to see that such a frame cannot be an affine frame, unless $\sigma = ++\cdots+$ (when we can use Ω_ν as the horizon) or $\sigma = --\cdots-$ (when we can use $\neg\Omega_\nu$). One possible choice for the horizon that agrees with these two special cases is the hyperplane $\langle\sigma_0, \sigma_1 - \sigma_0, \sigma_2 - \sigma_0, .. \sigma_\kappa - \sigma_0\rangle$. The corresponding map is

$$
M_{|s|} = \begin{bmatrix}
\left|1/s_0^0\right| & 0 & 0 & \cdots & 0 \\
-\left|1/s_0^0\right| & \left|1/s_0^1\right| & 0 & \cdots & 0 \\
-\left|1/s_0^0\right| & 0 & \left|1/s_0^2\right| & & \vdots \\
\vdots & \vdots & & \ddots & \vdots \\
-\left|1/s_0^0\right| & 0 & 0\cdots & & \left|1/s_0^\kappa\right|
\end{bmatrix}
\begin{bmatrix}
s_0^0 & s_1^0 & \cdots & \cdots & \cdots & s_\nu^0 \\
s_0^1 & s_1^1 & \cdots & \cdots & \cdots & s_\nu^1 \\
\vdots & \vdots & & & & \vdots \\
\vdots & \vdots & & & & \vdots \\
s_0^\kappa & s_1^\kappa & \cdots & \cdots & \cdots & s_\nu^\kappa
\end{bmatrix}
$$

$$
=
\begin{bmatrix}
\sigma_0 & \dfrac{s_1^0}{|s_0^0|} & \dfrac{s_2^0}{|s_0^0|} & \cdots & \dfrac{s_\nu^0}{|s_0^0|} \\[2ex]
\sigma_1 - \sigma_0 & \dfrac{s_1^1}{|s_0^1|} - \dfrac{s_1^0}{|s_0^0|} & \dfrac{s_2^1}{|s_0^1|} - \dfrac{s_2^0}{|s_0^0|} & \cdots & \dfrac{s_\nu^1}{|s_0^1|} - \dfrac{s_\nu^0}{|s_0^0|} \\[2ex]
\vdots & \vdots & \vdots & & \vdots \\[2ex]
\sigma_\kappa - \sigma_0 & \dfrac{s_1^\kappa}{|s_0^\kappa|} - \dfrac{s_1^0}{|s_0^0|} & \dfrac{s_2^\kappa}{|s_0^\kappa|} - \dfrac{s_2^0}{|s_0^0|} & \cdots & \dfrac{s_\nu^\kappa}{|s_0^\kappa|} - \dfrac{s_\nu^0}{|s_0^0|}
\end{bmatrix}
\tag{16}
$$

In particular, when the simplex s is a pair of points a, b, we get the "absolute" interpolation map

$$
M_{|a;b|} =
\begin{bmatrix}
\operatorname{sign} a_0 & \dfrac{a_1}{|a_0|} & \cdots & \dfrac{a_\nu}{|a_0|} \\[2ex]
\operatorname{sign}(b_0) - \operatorname{sign}(a_0) & \dfrac{b_1}{|b_0|} - \dfrac{a_1}{|a_0|} & \cdots & \dfrac{b_\nu}{|b_0|} - \dfrac{a_\nu}{|a_0|}
\end{bmatrix}
\tag{17}
$$

$$
=
\begin{bmatrix}
a_0\,|b_0| & a_1\,|b_0| & \cdots & a_\nu\,|b_0| \\[1ex]
b_0\,|a_0| - a_0\,|b_0| & b_1\,|a_0| - a_1\,|b_0| & \cdots & b_\nu\,|a_0| - a_\nu\,|b_0|
\end{bmatrix}
$$

The point c that is λ the way from a to b, in the "absolute" sense, is then

$$
c = [1, \lambda]\,M_{|a;b|}
$$

$$
= \Big[\, a_0\,|b_0| + \lambda\big(b_0\,|a_0| - a_0\,|b_0|\big),\ a_1\,|b_0| + \lambda\big(b_1\,|a_0| - a_1\,|b_0|\big),
$$
$$
\cdots,\ a_\nu\,|b_0| + \lambda\big(b_\nu\,|a_0| - a_\nu\,|b_0|\big) \,\Big]
\tag{18}
$$

$$
= [\, \alpha b_0 + \beta a_0,\ \alpha b_1 + \beta a_1,\ \alpha b_2 + \beta a_2,\ \ldots,\ \alpha b_\nu + \beta a_\nu \,]
\tag{19}
$$

where $\alpha = \lambda\,|a_0|$, and $\beta = (1 - \lambda)\,|b_0|$. That is, we scale each homogeneous tuple by the *positive* factors $|b_0|$ and $|a_0|$, so that their weights have the same *absolute* value but still the original signs, and then we interpolate the homogeneous coordinates in the ratio $\lambda : 1 - \lambda$. This formula agreees with formula (15) when both points are on the front range, and gives the intuitively correct result when both are on the back range. In either case, as λ goes from 0 to 1 the point c moves from a to b, with uniform "speed."

What happens now when a and b are on opposite ranges? It is not hard to see that as λ goes from 0 to 1 the point c will still traverse the segment ab, but not at a uniform rate: c speeds up as it moves away from a, reaches infinity when $\lambda = \frac{1}{2}$, and then slows down as it moves towards b on the opposite range. Note that this is a generalization of the "absolute" midpoint formula (4). Note also that both (4) and (19) give the null object $\mathbf{0}$ if a and b are both at infinity, or if a is the antipodal of b rotated $180°$ around the origin (that is, if $a = [-b_0, b_1, \ldots, b_\nu]$).

2.12. Measure of a simplex

In Cartesian geometry, the area of a triangle with vertices (x^0, y^0), (x^1, y^1), and (x^2, y^2) is given by

$$\frac{1}{6} \begin{vmatrix} 1 & x^0 & y^0 \\ 1 & x^1 & y^1 \\ 1 & x^2 & y^2 \end{vmatrix} \tag{20}$$

with the sign depending on the triangle's orientation. In general, the ν-dimensional measure of a simplex of $\mathbf{R}^\nu$ with vertices $x^i = (x_1^i, \ldots x_\nu^i)$ is given by

$$\frac{1}{(\nu+1)!} \begin{vmatrix} 1 & x_1^0 & \cdots & x_\nu^0 \\ 1 & x_1^1 & \cdots & x_\nu^1 \\ \vdots & \vdots & & \vdots \\ 1 & x_1^\nu & \cdots & x_\nu^\nu \end{vmatrix} \tag{21}$$

Combining this result with the homogeneous-to-Cartesian formulas we get an expression for the ν-dimensional measure of a simplex s of $\mathbf{A}_\nu$, namely

$$\frac{1}{n! \, s_0^0 s_0^1 \cdots s_0^\nu} \begin{vmatrix} s_0^0 & \cdots & s_\nu^0 \\ \vdots & & \vdots \\ s_0^\nu & \cdots & s_\nu^\nu \end{vmatrix} \tag{22}$$

For consistency, we may want to write this fraction in homogeneous form, as a point of $\mathbf{T}_1$,

$$\text{vol}(s) = \left[n! \cdot s_0^0 s_0^1 \cdots s_0^\nu, \begin{vmatrix} s_0^0 & \cdots & s_\nu^0 \\ \vdots & & \vdots \\ s_0^\nu & \cdots & s_\nu^\nu \end{vmatrix} \right] \tag{23}$$

Note that this formula is multilinear, which means

$$\text{vol}(\neg s^0, s^1, s^2, \ldots s^\nu) = \text{vol}(s^0, \neg s^1, s^2, \ldots s^\nu)$$
$$= \cdots = \text{vol}(s^0, s^1, s^2, \ldots \neg s^\nu) = \neg \, \text{vol}(s^0, s^1, s^2, \ldots s^\nu) \tag{24}$$

The orientation of s is given by the sign of the second coordinate of $\mathrm{vol}(s)$ only. An alternative, "absolute" formula is

$$\mathrm{vol}(s) = \left[\; n! \cdot \left|s_0^0 s_0^1 \cdots s_0^\nu\right| , \; \begin{vmatrix} s_0^0 & \cdots & s_\nu^0 \\ \vdots & & \vdots \\ s_0^\nu & \cdots & s_\nu^\nu \end{vmatrix} \;\right] \tag{25}$$

which is always a point of the front range of $\mathbf{T}_1$ (or infinite), and whose numerical sign gives the orientation of s.

  Since affine maps preserve the ratio of measures, we can use formula (23) or (25) to compute the ratio between the measures of two simplices s, t contained in the same κ-dimensional affine subspace A of $\mathbf{A}_\nu$. We only need to compute an affine map M that takes A to $\mathbf{A}_\kappa$, compute the measures of $M(s)$ and $M(t)$ in $\mathbf{A}_\kappa$, and take their ratio. We can compute this ratio even if s and t lie in distinct but parallel affine subspaces A, B of $\mathbf{T}_\nu$: it suffices to compute a translation T that takes B to A, and proceed as above with s and $T(t)$.

  This is the best we can do within affine geometry. To compute the absolute simplex measure in a proper subspace of $\mathbf{A}_\nu$, or the simplex measure ratio between non-parallel subspaces, we need non-affine concepts such as congruence and distance-preserving maps, which we will intriduce in chapter 17.

Chapter 16
Vector algebra

The linear vector space $\mathbf{R}^\nu$ can be emulated in a ν-dimensional affine space A by designating a fixed point o on the front range of A to represent the origin of $\mathbf{R}^\nu$. By this device, we can encode any vector v of $\mathbf{R}^\nu$ as a point $\dot{v}$ on the front range of A. The addition of two vectors u, v can be carried out by construcing the parallelogram with corners at o, $\dot{u}$, and $\dot{v}$. The product of a vector v by a scalar λ can be done by affine interpolation (or extrapolation) between o and $\dot{v}$, in the ratio $\lambda : 1 - \lambda$.

By embedding $\mathbf{R}^\nu$ in a two-sided space, we gain the ability to handle infinite-length vectors, which are useful in some geometric algorithms. As a bonus, we get also a second copy of $\mathbf{R}^\nu$, namely the points on the back range of A. While this feature may be of limited utility, it is generally harmless: we can "turn it off' by simply ignoring the distinction between v and $\neg v$.

1. Two-sided vector spaces

Definition 1. A *two-sided vector space* is a triple $V = (S, h, o)$ where S is a two-sided space, h (the *horizon*) is a hyperplane of S, and o (the *origin*) is a point of S on the positive side of h.

In particular, we can take $S = \mathbf{T}_\nu$, $h = \Omega_\nu$, and $o = O_\nu = [1, 0, 0, ..]$. These choices define the *canonical two-sided vector space* $\mathbf{V}_\nu = (\mathbf{T}_\nu, \Omega_\nu, O_\nu)$. With these choices, a vector $(x_1, x_2, .. x_\nu)$ of $\mathbf{R}^\nu$ is represented by the point $(1, x_1, x_2, .. x_\nu)$ of $\mathbf{V}_\nu$ — which, of course, is merely the familiar Cartesian-to-homogeneous transformation.

A subspace of a two-sided vector space $V = (S, h, o)$ is any triple (f, g, o) where f is a flat of S that contains o, and $g = f \wedge h$. That is, (f, g) is an affine subspace of (S, h) that includes the point o. In particular, $(\neg S, \neg h, o)$ is a vector subspace of $V = (S, h, o)$, denoted by $\neg V$. The two spaces $(S, \neg h, \neg o)$ and $(\neg S, h, \neg o)$ are subspaces of each other, but not of V. Observe that the other four variants $(S, h, \neg o)$, $(S, \neg h, o)$, $(\neg S, h, o)$, and $(\neg S, \neg h, \neg o)$, are not vector spaces, according to the definition.

2. Translations

In general terms, a translation is a projective map between two co-parallel subspaces of an affine space $A = (S, h)$ that preserves the direction of every subflat. Obviously, a translation is a special case of affine map, one that maps every point of h to itself, and not just the hyperplane h to itself. A translation in a vector space $V = (S, h, o)$ can be uniquely characterized by the image of the origin o. This establishes a correspondence between translations and the points (vectors) on the front range of V. In particular, the translation of $\mathbf{V}_\nu$ that takes the origin O_ν to the point $x = [x_0, .. x_\nu]$ is

$$
\begin{bmatrix}
x_0 & x_1 & x_2 & \cdots & x_\nu \\
0 & x_0 & 0 & \cdots & 0 \\
0 & 0 & x_0 & & 0 \\
\vdots & \vdots & & \ddots & \vdots \\
0 & 0 & 0 & \cdots & x_0
\end{bmatrix}
\tag{1}
$$

Conversely, any matrix of this form with $x_0 > 0$ defines a translation. If x_0 is negative (meaning the point x is on the back range) then the corresponding translation (1) is an affine map from $\mathbf{A}_\nu$ to $\neg^n(\mathbf{T}_\nu, \neg\Omega_\nu)$, that swaps the front and back ranges.

3. Vector algebra

3.1. Vector addition

If we equate the vectors of a two-sided vector space V with its translation maps, the sum of two vectors $[x]$ and $[y]$ will correspond to the composition of the respective translations, namely

$$
\begin{bmatrix}
x_0 & x_1 & x_2 & \cdots & x_\nu \\
0 & x_0 & 0 & \cdots & 0 \\
0 & 0 & x_0 & & 0 \\
\vdots & \vdots & & \ddots & \vdots \\
0 & 0 & 0 & \cdots & x_0
\end{bmatrix}
\begin{bmatrix}
y_0 & y_1 & y_2 & \cdots & y_\nu \\
0 & y_0 & 0 & \cdots & 0 \\
0 & 0 & y_0 & & 0 \\
\vdots & \vdots & & \ddots & \vdots \\
0 & 0 & 0 & \cdots & y_0
\end{bmatrix}
$$

$$
= \begin{bmatrix}
x_0 y_0 & x_0 y_1 + y_0 x_1 & x_0 y_2 + y_0 x_2 & \cdots & x_0 y_\nu + y_0 x_\nu \\
0 & x_0 y_0 & 0 & \cdots & 0 \\
0 & 0 & x_0 y_0 & & 0 \\
\vdots & \vdots & & \ddots & \vdots \\
0 & 0 & 0 & \cdots & x_0 y_0
\end{bmatrix}
$$

Therefore, the sum of vectors $[x]$ and $[y]$ is

$$
\begin{aligned}
&[x_0, \ldots x_\nu] + [y_0, \ldots y_\nu] \\
&= \left[x_0 y_0, \ x_0 y_1 + y_0 x_1, \ x_0 y_2 + y_0 x_2, \ \ldots, \ x_0 y_\nu + y_0 x_\nu \right]
\end{aligned}
\tag{2}
$$

Another way of deriving this formula is to apply to one of the vectors (viewed as a point) the translation map associated with the other.

Note that in the derivation of formula (2) we assumed that x_0 and y_0 are positive, that is, the two vectors are points of the front range. If only one of them is a point at infinity, the sum will be that same point; if both x and y are at infinity, the sum is the undefined point $\mathbf{0}$. As for points on th eback range, observe that formula (2) is multilinear, and hence a "natural" formula in the sense discussed before: we have $(\neg x) + y = x + (\neg y) = \neg(x + y)$ for all x, y. It follows that $x + y$ is a front vector if x and y are on the same range, and is a back vector if they are on different ranges. Note that adding the front origin $O_\nu = [1, 0, 0, \ldots 0]$ to a vector x produces x itself, whereas adding the back origin $\neg O_\nu = [-1, 0, 0, \ldots 0]$ returns the antipodal vector $\neg x$ (the vector with the same Cartesian coordinates on the opposite range of $\mathbf{V}_\nu$).

3.2. Vector negation

The inverse of the translation map (1) is given by the matrix

$$
\begin{bmatrix}
x_0 & -x_1 & -x_2 & \cdots & -x_\nu \\
0 & x_0 & 0 & \cdots & 0 \\
0 & 0 & x_0 & & 0 \\
\vdots & \vdots & & \ddots & \vdots \\
0 & 0 & 0 & \cdots & x_0
\end{bmatrix}
$$

This gives a formula for the additive inverse of a vector,

$$
-[x_0, \ldots x_\nu] = [x_0, -x_1, -x_2, \ldots -x_\nu]
\tag{3}
$$

Indeed, we have

$$x + (-x) = \left[\, x_0 x_0 \,,\, x_0 x_1 - x_0 x_1 \,,\, x_0 x_2 - x_0 x_2 \,,\, \ldots \,,\, x_0 x_\nu + x_0 x_\nu \,\right]$$
$$= [(x_0)^2, 0, 0, \ldots, 0]$$

which is the front origin O_ν if x is a proper point, and the null object $\mathbf{0}$ if x is at infinity. The difference of two vectors is then

$$[x_0, \ldots x_\nu] - [y_0, \ldots y_\nu]$$
$$= \left[\, x_0 y_0 \,,\, y_0 x_1 - x_0 y_1 \,,\, y_0 x_2 - x_0 y_2 \,,\, \ldots \,,\, y_0 x_\nu - x_0 y_\nu \,\right] \tag{4}$$

3.3. "Absolute" vector addition

The following is an "absolute" alternative for the "natural" vector addition formula (2):

$$[x_0, \ldots x_\nu] + [y_0, \ldots y_\nu]$$
$$= \left[\, \frac{1}{2}(x_0 |y_0| + y_0 |x_0|) \,,\, |x_0| y_1 + |y_0| x_1 \,,\, \ldots \,,\, |x_0| y_\nu + |y_0| x_\nu \,\right] \tag{5}$$

If x and y are both on the front range or both on the back range, formula (5) puts $x+y$ on that same range; otherwise it returns a point at infinity in the direction of $(\neg x) + y$ or $x + (\neg y)$, depending on whether x is on the front or back range, respectively. However, if $x = \neg(-y)$ the result is the null object $\mathbf{0}$.

3.4. Multiplication by a scalar

The *scalar multiplication* of a point $x = [x_0, \ldots]$ by a real number β is given by $\beta \cdot x = [x_0, \beta x_1, \beta x_2, \ldots]$. Equivalently, we can use the formula $\beta x = [x_0/\beta, x_1, x_2, \ldots]$ if $\beta > 0$, or $\beta x = [-x_0/\beta, -x_1, -x_2, \ldots]$ if $\beta < 0$. Note that $-1 \cdot x = [x_0, -x_1, -x_2, \ldots]$ is the additive inverse $-x$ of x, as per formula (3).

The proper way to view these formulas is to imagine the scalars as elements of $\mathbf{T}_1$, the two-sided line. Then the scalar multiplication is given by

$$[\beta_0, \beta_1] \cdot [x_0, x_1, x_2, \ldots] = [\beta_0 x_0, \beta_1 x_1, \beta_1 x_2, \ldots]$$

The range of $\mathbf{T}_\nu$ on which the product $\beta \cdot x$ lies is determined by the product of the signs of their weights β_0 and x_0. So, for example, multiplication by $[3, -2]$ (which lies on the front range of $\mathbf{T}_1$) produces the vector $(-\frac{2}{3}) \cdot x$ on the same range as x, whereas multiplication by $[-3, 2]$ gives the antipodal of the above vector.

4. The two-sided real line

When $\nu = 1$, the operations of addition and multiplication by a scalar defined above become addition and multiplication of two elements of $\mathbf{T}_1$:

$$[x_0, x_1] + [y_0, y_1] = [x_0 y_0, x_0 y_1 + y_0 x_1]$$
$$[x_0, x_1] \cdot [y_0, y_1] = [x_0 y_0, x_1 y_1] \tag{6}$$

Subtraction and division are given by the formulas

$$-[x_0, x_1] = [x_0, -x_1]$$
$$1/[x_0, x_1] = [x_1, x_0] \tag{7}$$

Note however that

$$x + (-x) = [x_0, x_1] + [x_0, -x_1] = [(x_0)^2, 0]$$

which is normally the front origin, but is the null object $\mathbf{0}$ when x is at infinity (i.e. $x_0 = 0$). Also,

$$x \cdot (1/x) = [x_0, x_1] \cdot [x_1, x_0]) = [x_0 x_1, x_0 x_1]$$

which is $[1, 1]$ if x is positive, $[-1, -1] = \neg[1, 1]$ if x is negative, and $[0, 0] = \mathbf{0}$ if x is infinite or zero (i.e., $x = [0, \pm 1]$ or $[\pm 1, 0]$). With these caveats, formulas (6) and (7) allow us to do arithmetic on $\mathbf{T}_1$ as if it were a two-sided version of the real line.

5. Linear maps

The linear maps of $\mathbf{R}^\nu$ are simulated in $\mathbf{T}_\nu$ by projective maps that keep both Ω_ν and the origin fixed. They have the form

$$
\begin{bmatrix}
a_{00} & 0 & 0 & \cdots & 0 \\
0 & a_{11} & a_{12} & \cdots & a_{1\nu} \\
0 & a_{21} & a_{22} & \cdots & a_{2\nu} \\
\vdots & \vdots & \vdots & & \vdots \\
0 & a_{\nu 1} & a_{\nu 2} & \cdots & a_{\nu\nu}
\end{bmatrix}
$$

where a_{00} is positive, and corresponds to the linear map of $\mathbf{R}^\nu$ to $\mathbf{R}^\nu$ whose coefficient matrix consists of the ratios a_{ij}/a_{00} for $i,j \in \{1..\nu\}$. In particular,

$$
\begin{bmatrix}
x_0 & 0 & 0 & \cdots & 0 \\
0 & x_1 & 0 & \cdots & 0 \\
0 & 0 & x_2 & & 0 \\
\vdots & \vdots & & \ddots & \vdots \\
0 & 0 & 0 & \cdots & x_\nu
\end{bmatrix}
$$

is the matrix of a *scaling map*, whose effect is to multiply the ith Cartesian coordinate by the ratio x_i/x_0. A convenient way to specify such a map (for example, to a graphics package, or to a procedure that builds scaling maps) is to give the point $[x_0, x_1, .., x_\nu]$, which is the image of the standard unit point $\mathbf{u} = [1, 1, .. 1]$ by the desired map. When $x_1 = x_2 = \cdots = x_\nu$ we get a *uniform scaling*, whose effect is to scale all vectors by the same ratio x_1/x_0. Again, for consistency we may want to think of this ratio as the two-sided fraction $[x_0, x_1]$.

Chapter 17
Euclidean geometry on the two-sided plane

Euclidean geometry can be emulated in the two-sided space $\mathbf{T}_\nu$ by treating the two ranges of the straight model as copies of the ν-dimensional Euclidean space, and defining perpendicularity, congruence, angular measure, and other Euclidean concepts by reference to the standard formulas of Cartesian geometry. In this chapter, I define the canonical *two-sided Euclidean space* following this approach.

Once this canonical representative exists, the notion of abstract two-sided Euclidean space can be defined, as usual, as any structure that is isomorphic to the canonical one. In order to do this, however, we must locate some "fundamental notions" or "distinguished objects" in the space, from which we can derive all Euclidean concepts, such as perpendicularity and distance, by purely projective tools. That is, we need something that can play in Euclidean geometry the same role that the horizon hyperplane plays in affine geometry. The "fundamental objects" I have chosen to use are a distinguished *horizon* hyperplane h, and a polarity-like relation defined on the points of h. In the canonical space these are by definition Ω_ν and the standard polarity $\perp$.

As usual, when adapting the formulas of Cartesian geometry to the two-sided model, we need to make a few choices about the signs of coordinates. These and other details are discussed below.

1. Perpendicularity

First of all, let's define what it means for two flats of $\mathbf{T}_\nu$ to be perpendicular. We want this condition to be true if and only if their front parts are perpendicular in the Cartesian sense. First, we need some auxiliary definitions:

1.1. Orthogonal directions

Let a, b be two directions of $\mathbf{T}_\nu$, that is, two subflats of Ω_ν. By definition, a and b are *orthogonal* if they are polar to each other. In terms of the analytic model, two points $[x], [y]$ of Ω_ν are orthogonal if and only if their homogeneous coordinates satisfy $x_1 y_1 + \cdots + x_\nu y_\nu = 0$.

1.2. Normals

Let a be a proper flat of $\mathbf{T}_\nu$. Its *normal direction* (or simply its *normal*) is the improper flat norm(a) satisfying

$$\begin{aligned}
\text{dir}(a) \perp \text{norm}(a) \\
\text{dir}(a) \vee \text{norm}(a) = \Omega_\nu
\end{aligned} \tag{1}$$

In other words, norm(a) is $\Omega \lceil \text{dir}(a)$, the right polar complement of a relative to Ω. The points of norm(a) are all points at infinity that are orthogonal to dir(a), It follows from this definition that

$$\begin{aligned}
\text{norm}(a) = (O \vee \text{dir}(a))^\vdash \\
a \vee \text{norm}(a) = \mathbf{T}_\nu
\end{aligned} \tag{2}$$

In classic Euclidean geometry, the normal direction of a hyperplane is usually encoded as a unit vector, and its orientation is not specified. The two-sided definition, interpreted in the straight model, basically agrees with the classical one, except that it fully specifies the orientation of the normal, and encodes it as a point at infinity.

For example, the normal of a line l of $\mathbf{T}_2$ is a point at infinity in the direction 90° counterclockwise from l, as seen from the front range. See figure 1.

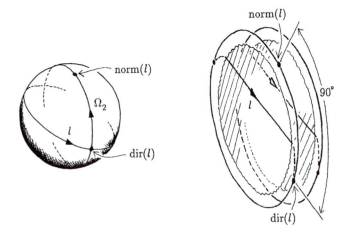

Figure 1. The normal of a line in $\mathbf{T}_2$.

The normal norm(π) of a proper plane π of $\mathbf{T}_3$ is a point at infinity whose direction is perpendicular to π (in the straight model of $\mathbf{T}_3$). The orientation of norm(π) is derived from that of π by the right-hand rule, as shown in figure 2.

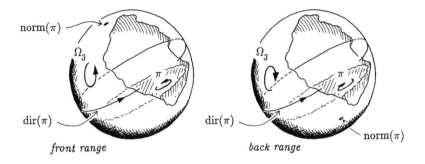

Figure 2. The normal of a plane in $\mathbf{T}_3$.

The normal of a line l in $\mathbf{T}_3$ is a line at infinity, which is the direction of any proper plane that is perpendicular to l, in the usual sense. The line l and its normal are positively oriented in $\mathbf{T}_3$. See figure 3.

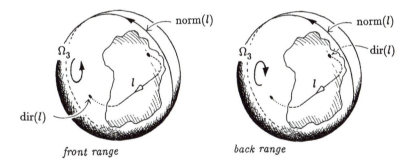

Figure 3. The normal of a line in $\mathbf{T}_3$.

Observe that $\mathrm{norm}(\pi)$ is on the *negative* side of π, but $\mathrm{norm}(l)$ is on the positive side of l. In general,

$$a \diamond \mathrm{norm}(a) \;=\; +1$$

$$\mathrm{norm}(a) \diamond a \;=\; (-1)^{\mathrm{rank}(a)\,\mathrm{corank}(a)}$$

In particular, $\mathrm{norm}(h) \diamond h = (-1)^{\nu}$ for any hyperplane h of $\mathbf{T}_{\nu}$.

1.3. Perpendicular flats

In general, I will say that two proper flats a, b are *perpendicular* if $\mathrm{dir}(a) \subseteq$ $\mathrm{norm}(b)$, or $\mathrm{norm}(a) \subseteq \mathrm{dir}(b)$. For example, two proper lines l, m of $\mathbf{T}_\nu$ are perpendicular if and only if their directions are two polar points of Ω. In the straight model, this condition means there is a plane that contains one line and is perpendicular to the other. Note that two perpendicular lines of $\mathbf{T}_3$ may or may not intersect. See figure 4.

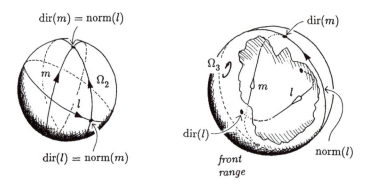

Figure 4. Perpendicular lines of $\mathbf{T}_2$ and $\mathbf{T}_3$.

In $\mathbf{T}_3$, a line l is perpendicular to a plane π if and only if $\mathrm{norm}(\pi)$ is $\mathrm{dir}(l)$ or $\neg\,\mathrm{dir}(l)$. In the straight model, this definition agrees with the usual one. See figure 5.

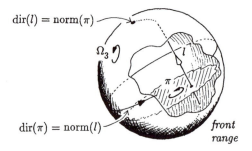

Figure 5. Perpendicularity between a line and a plane of $\mathbf{T}_3$.

Two planes of $\mathbf{T}_3$ are perpendicular if and only if their normals are orthogonal to each other; that is, the directions of the two planes are two orthogonal great circles of the celestial sphere Ω_3. See figure 6.

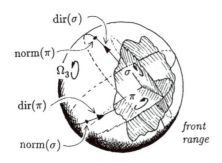

Figure 6. Perpendicular planes of $\mathbf{T}_3$.

Since normals and directions are complementary with respect to Ω, we have $\mathrm{dir}(a) \subseteq \mathrm{norm}(b) \Leftrightarrow \mathrm{dir}(b) \subseteq \mathrm{norm}(a)$. It follows that perpendicularity, like parallelism, is a symmetric relation: a is perpendicular to b if and only if b is perpendicular to a.

Note that if a is perpendicular to b, and b to c, we cannot in general conclude that a is parallel to c. For a counterexample, let a, b, and c be the Cartesian axes of $\mathbf{T}_3$. However, the conclusion $a \parallel b$ is legitimate for some rank combinations, namely when $\mathrm{rank}(a) \geq 1 + \mathrm{corank}(b) \geq \mathrm{rank}(c)$, or $\mathrm{rank}(a) \leq 1 + \mathrm{corank}(b) \leq \mathrm{rank}(c)$. Two common instances are when a, b, c are lines of $\mathbf{T}_2$, or a and c are lines and b is a plane of $\mathbf{T}_3$.

Formulas for many familiar constructions of Euclidean geometry follow readily from the definitions above. For example, the flat of maximum rank that is perpendicular to a proper flat a and passes through a proper point p is given by the formula $p \vee \mathrm{norm}(a)$. In particular, the perpendicular bisector of a segment pq is given by $m \vee \mathrm{norm}(p \vee q))$, where m is the segment's midpoint. The perpendicular projection onto a flat a is the map $x \mapsto (x \vee \mathrm{norm}(a)) \wedge a$. And so on.

2. Two-sided Euclidean spaces

All concepts of Euclidean geometry can be derived from the notion of perpendicularity. Recall that we defined perpendicularity in $\mathbf{T}_\nu$ in terms of affine operations and the polarity predicate $\perp$, restricted to the points of Ω. So, we can define perpendicularity and all other Euclidean concepts in an arbitrary two-sided space by letting any fixed hyperplane h play the role of Ω, and any suitable relation η on the subflats of h play the role of the polarity predicate.

Definition 1. The *canonical two-sided Euclidean space* of dimension ν is the triple $\mathbf{E}_\nu = (\mathbf{T}_\nu, \Omega_\nu, \perp_\Omega)$, where $\perp_\Omega$ is the standard polarity relation restricted to subflats of Ω_ν.

Definition 2. A *two-sided Euclidean space* is a triple (S, h, ρ) isomorphic to the canonical space $\mathbf{E}_\nu$, for some ν.

Here "isomorphic" means there is projective map η from S to $\mathbf{T}_\nu$ such that $h\eta = \Omega_\nu$, and $x \rho y \Leftrightarrow (x\eta) \perp (y\eta)$, for all points x, y of h. Note that a two-sided Euclidean space is a two-sided affine space with the extra structure given by the orthogonality predicate ρ.

Definition 3. A *subspace* of a Euclidean space (S, h, ρ) is a triple (T, g, σ) where T is a projective subspace of S, g is $T \wedge_S h$, and σ is the restriction of the relation ρ to the subflats of g.

As in the affine case, there are three other spaces with the same point set as $\mathbf{E}_\nu$ and the same points at infinity, namely $(\pm\mathbf{T}_\nu, \pm\Omega, \perp_\Omega)$. Of these only $(\neg\mathbf{T}_\nu, \neg\Omega, \perp_\Omega)$ is a subspace of $\mathbf{E}_\nu$, denoted by $\neg\mathbf{E}_\nu$.

3. Euclidean maps

In classical geometry a Euclidean map can be defined as a map that preserves congruence: two segments have equal length if and only if their images have equal length. Examples are translations, rigid rotations, and uniform scalings. Euclidean maps also preserve perpendicularity, and indeed they are the only maps that do so. Therefore, we can use this property to define Euclidean maps of $\mathbf{T}_\nu$:

Definition 4. A *Euclidean map* or *similarity* is an isomorphism between two Euclidean spaces.

Here, and isomorphism from the two-sided Euclidean space $E = (S, h, \rho)$ to the space $F = (T, g, \sigma)$ is a projective map φ from S to T such that $h\varphi = g$, and $(x\varphi) \sigma (y\varphi) \Leftrightarrow x \rho y$ for all $x, y \subseteq h$. In other words, an Euclidean map is an affine map that takes orthogonal directions to orthogonal directions.

Two trivial examples of Euclidean maps from $\mathbf{T}_\nu$ to itself are the translations and uniform scalings defined in chapter 15: those maps keep all points of Ω_ν fixed, and therefore trivially preserve the polarity relation among those points.

3.1. Analytic characterization

A Euclidean map of $\mathbf{T}_\nu$ has a matrix of the form

$$
M = \begin{bmatrix}
m_0^0 & m_1^0 & m_2^0 & \cdots & m_\nu^0 \\
0 & m_1^1 & m_2^1 & \cdots & m_\nu^1 \\
0 & m_1^2 & m_2^2 & \cdots & m_\nu^2 \\
\vdots & \vdots & \vdots & & \vdots \\
0 & m_1^\nu & m_2^\nu & \cdots & m_\nu^\nu
\end{bmatrix}
\tag{3}
$$

where m_0^0 and its cofactor are positive, and rows 1 through ν are orthogonal vectors with identical length λ. That is,

$$
\sum_{1 \le k \le \nu} m_k^i m_k^j = 0 \quad \text{and} \quad \sum_{1 \le k \le \nu} (m_k^i)^2 = \lambda^2 \qquad \text{for } 1 \le i < j \le \nu.
\tag{4}
$$

It is easy to check that any map of this form takes $\varUpsilon_\nu$ to $\varUpsilon_\nu$, Ω to Ω, and preserves polarities on Ω. Conversely, it is easy to show that any map that preserves the polarity relations between all improper points of the form $[\mathbf{e}^i]$ and $[\mathbf{e}^i \pm \mathbf{e}^j]$, for $1 \le i < j \le n$, must be of this form.

Now consider a map M that satisfies (3) and (4) but where m_0^0 and/or its cofactor are negative. Such a map still takes improper points to improper points and preserves their polarities, but reverses the orientations of $\varUpsilon_\nu$ and/or Ω_ν. That means M is a map from $\mathbf{E}_\nu$ to one of the four Euclidean spaces $(\pm\mathbf{T}_\nu, \pm\Omega, \perp_\Omega)$. For example, consider the vector negation map N, which sends every vector of $\mathbf{V}_\nu$ to its additive inverse, and the antipode map A, which sends every point to its antipode:

$$
N = \begin{bmatrix}
1 & & & 0 \\
& -1 & & \\
& & \ddots & \\
0 & & & -1
\end{bmatrix}
\qquad
A = \begin{bmatrix}
-1 & & & 0 \\
& -1 & & \\
& & \ddots & \\
0 & & & -1
\end{bmatrix}
$$

Note that

$$
\begin{aligned}
\varUpsilon_\nu N &= \neg^\nu \varUpsilon_\nu & \varUpsilon_\nu A &= \neg^n \varUpsilon_\nu & \varUpsilon_\nu A N &= \neg \varUpsilon_\nu \\
\Omega_\nu N &= \neg^\nu \Omega_\nu & \Omega_\nu A &= \neg^\nu \Omega_\nu & \Omega_\nu A N &= \Omega_\nu
\end{aligned}
\tag{5}
$$

So, N is a map from $\mathbf{E}_\nu$ to $\neg^\nu \mathbf{E}_\nu$, and A is a map from $\mathbf{E}_\nu$ to $\neg^\nu(\neg\mathbf{T}_\nu, \Omega, \perp_\Omega)$.

In what follows, an "Euclidean map of a space $E = (S, h, \rho)$" means a Euclidean map from a space E to any of its four related spaces $(\pm S, \pm h, \rho)$.

3.2. Isometries

An *isometry* is a Euclidean map that preserves volumes. Analytically, this condition means that the quantity λ^2 in formula (4) is equal to $(m_0^0)^2$. Translations are obviously isometries, and so are the maps N and A defined above. It is easy to show that every Euclidean map is the composition of an isometry with a uniform scaling. More precisely, a map of $\mathbf{E}_\nu$ that enlarges the volume of every simplex by a factor of τ is the product of an isometry of $\mathbf{E}_\nu$ and a uniform scaling by $|\tau|^{1/n}$.

3.3. Rotations

I will define a *rotation* or *rigid motion* of a Euclidean space as an isometry of the space to itself, that is, an isometry that preserves the orientations of the universe and of the horizon. A typical rotation of $\mathbf{E}_\nu$ is any orthonormal transformation of $\mathbf{R}^\nu$ with determinant $+1$, applied to both ranges of the straight model. The rotations of $\mathbf{E}_\nu$ include also the positive translations of $\mathbf{A}_\nu$, and various helical (screw-like) motions.

Rotations are closed under composition and inverse. Every rotation can be written as the product of a translation and a rotation that doesn't move the origin. For any pair of distinct axes i, j in $\{1, \nu\}$, and any angle θ, the map

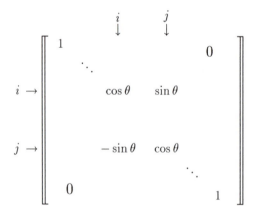

rotates each range of the straight model by the angle θ around a $(\nu-2)$-dimensional "axis," the hyperline $(\mathbf{e}^i \vee \mathbf{e}^j)^\vdash$; that is, it rotates $\mathbf{e}^i$ towards $\mathbf{e}^j$, and $\mathbf{e}^j$ towards $-\mathbf{e}^i$, while leaving all other cardinal points $\mathbf{e}^k$ fixed. (If $\theta = 90°$, the map takes $\mathbf{e}^i$ to $\mathbf{e}^j$, and $\mathbf{e}^j$ to $-\mathbf{e}^i$). Let's call the maps of this form *principal rotations* of $\mathbf{E}_\nu$. A rotation of $\mathbf{E}_\nu$ that does not move the origin can be written as the product of $\nu(\nu-1)/2$ principal rotations.

3.4. Reflection across a hyperplane

Isometries that map the universe S to $\neg S$ are called *reflections*. An important example is reflection across a proper hyperplane $h = \langle h \rangle = \langle h^0, .. h^\nu \rangle$ of $\mathbf{E}_\nu$. Let g be the hyperplane co-parallel to h that passes through O, that is, $g = \langle g \rangle$ where $g = (0, h^1, .. h^\nu)$. Then the (*Euclidean*) *reflection across* h is by definition the map

$$R_h = [\![(g h^{\mathrm{tr}})I - 2h^{\mathrm{tr}}g]\!]$$

Expanded, this formula becomes

$$R_h = \left[\!\!\left[\begin{array}{ccccc} \lambda^2 & -2h^0h^1 & -2h^0h^2 & \cdots & -2h^0h^\nu \\ 0 & \lambda^2 - 2h^1h^1 & -2h^1h^2 & \cdots & -2h^1h^\nu \\ \vdots & -2h^2h^1 & \lambda^2 - 2h^2h^2 & & \vdots \\ \vdots & \vdots & & \ddots & \\ 0 & -2h^\nu h^1 & \cdots & & \lambda^2 - 2h^\nu h^\nu \end{array}\right]\!\!\right] \qquad (6)$$

where $\lambda^2 = g h^{\mathrm{tr}} = \sum_{k=1}^{\nu} h^k h^k$. It is straightforward to check that R_h is an isometry, that it keeps every point on h fixed, and that it swaps the point $\mathrm{norm}(h) = [-g] = [0, -h^1, .. -h^\nu]$ with its antipode. See figure 7. Note that all coefficients of R_h

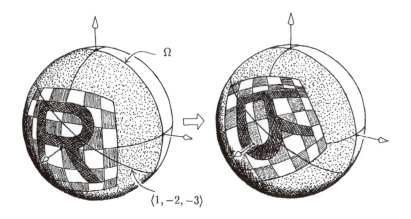

Figure 7. Reflection of $\mathbf{T}_2$ across the line $\langle 1, -2, -3 \rangle$.

in formula (6) are homogeneous second-degree polynomials in the coefficients of h. This means the orientation of h is irrelevant, i.e. $R_h = R_{\neg h}$.

3.5. Reflection across Ω

Another example of reflection is the map

$$R_\Omega = AN = \begin{bmatrix} -1 & & & 0 \\ & 1 & & \\ & & \ddots & \\ 0 & & & 1 \end{bmatrix}$$

which in the straight model sends the point $(x_1, .. x_\nu)$ of the front range to point $(-x_1, .. -x_\nu)$ of the back, and vice versa. That is, it sends vector x of the two-sided vector space $\mathbf{V}_\nu$ to the antipodal of its additive inverse, the vector $\neg(-x)$.

In the spherical model, this map simply mirrors the unit sphere across the plane $x_0 = 0$ of $\mathbf{R}^n$. We may call this map *reflection across* Ω. Note that it takes $\mathbf{E}_\nu$ to the space $(\neg\mathbf{T}_\nu, \Omega, \perp_\Omega)$. See figure 8.

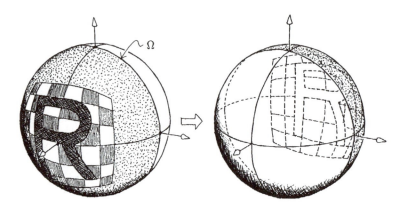

Figure 8. Reflection of $\mathbf{T}_2$ across Ω.

The product of two reflections is a rotation. For example, the product $R_h R_g$, where h and g are parallel hyperplanes, is equivalent to a translation by twice the displacement from h to g. If h and g intersect in a proper hyperline l, the product $R_h R_g$ is (in the straight model) a rotation of each range around the "axis" l, by twice the angle between h and g. In particular, reflection across two perpendicular hyperplanes gives a rotation of 180° around their common hyperline l: every point x is mapped to the point y such that l is perpendicular to the segment xy, and passes through its midpoint.

This statement is true in general for the result of reflecting $\mathbf{T}_\nu$ across k pairwise perpendicular hyperplanes, with l replaced by the flat set of co-rank k that is their common intersection. It is customary to call the resulting map R_l the *reflection across l*, even when it is actually a rotation (k even). For example, by reflecting across all the proper main hyperplanes $\langle \mathbf{e}^1 \rangle$, $\langle \mathbf{e}^2 \rangle$, ..., $\langle \mathbf{e}^\nu \rangle$, we get the *reflection across the origin*,

$$
R_O = \begin{bmatrix} 1 & & & 0 \\ & -1 & & \\ & & \ddots & \\ 0 & & & -1 \end{bmatrix}
$$

It can be shown that any isometry of $\mathbf{E}_\nu$ is the product of at most $\nu + 1$ reflections across proper hyperplanes, and at most one instance of R_Ω. The isometry is a rotation or a reflection depending on whether the total number of terms is even or odd, and it swaps the back and front ranges if and only if it includes R_Ω.

4. Length and distance

Euclidean maps turn out to preserve ratios of distances as well as perpendicularity. To make this statement meaningful, we must define the ratio of distances in an arbitrary two-sided Euclidean space. To achieve this end, I will introduce first a numerical distance function for points of $\mathbf{E}_\nu$.

4.1. Two-sided length of a vector

Let $[x]$ be a finite vector on the front range of $\mathbf{V}_\nu$. The length of $[x]$, viewed as a vector in the straight model, is given by the familiar formula

$$
\sqrt{\left(\frac{x_1}{x_0}\right)^2 + \cdots \left(\frac{x_\nu}{x_0}\right)^2} = \frac{1}{x_0}\sqrt{(x_1)^2 + \cdots (x_\nu)^2}
$$

or, as a two-sided fraction,

$$
\text{len}(x) = [\, x_0, \sqrt{(x_1)^2 + \cdots (x_\nu)^2}\,] \tag{7}
$$

I will take this formula as defining the *two-sided length* of the vector $[x]$.

Note that the weight x_0 is the only coordinate of x whose sign affects the two-sided length of x. Specifically, the length of a front vector is a positive front number, and that of a back vector is a negative back number.

For example,

$$\text{len}\,[+1,+1,+1] = [+1,\sqrt{2}] = \sqrt{2}$$
$$\text{len}\,[+1,-1,-1] = [+1,\sqrt{2}] = \sqrt{2}$$
$$\text{len}\,[-1,+1,+1] = [-1,\sqrt{2}] = \neg(-\sqrt{2})$$
$$\text{len}\,[-1,-1,-1] = [-1,\sqrt{2}] = \neg(-\sqrt{2})$$

4.2. Two-sided distance between two points

I also define the *two-sided distance* between two points of the front range of $\mathbf{T}_\nu$ is being the two-sided length of their difference, computed as if they were elements of $\mathbf{V}_\nu$. From (7) and from the formula for vector difference (4, chapter 16), the two-sided distance between points $[x]$ and $[y]$ is

$$\text{dist}(x,y) = \text{len}(x - y)$$

$$= \text{len}\big[\, x_0 y_0,\ y_0 x_1 - x_0 y_1,\ y_0 x_2 - x_0 y_2,\ \ldots,\ y_0 x_\nu - x_0 y_\nu \,\big] \qquad (8)$$

$$= \big[\, x_0 y_0,\ \sqrt{(y_0 x_1 - x_0 y_1)^2 + \cdots + (y_0 x_\nu - x_0 y_\nu)^2}\,\big]$$

Note that $\text{dist}(x,y) = \text{dist}(y,x)$, and $\text{dist}(x,\neg y) = \text{dist}(\neg x,y) = \neg(-\text{dist}(x,y))$ for all x,y. Also, $\text{dist}(x,x) = 0$, $\text{dist}(x,\neg x) = \neg 0$ for any proper point x. From the properties of vector sum in $\mathbf{V}_2$ it follows that the distance as defined above is a front positive number if x and y are on the same range, a back negative number if they are in opposite ranges, plus infinity if one of them is at infinity, and undefined if both are at infinity.

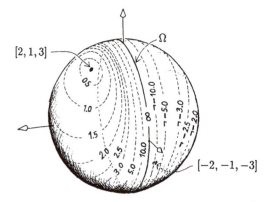

Figure 9. Distances from $[2,1,3]$ on the two-sided plane.

More precisely, the distance increases monotonically from 0 to ∞ as y moves away from x on the same range, and then increases monotonically from $\neg\infty$ to $\neg 0$ as y moves on the opposite range from Ω towards $\neg x$. See figure 9. In the straight model, the curves of equal distance in figure 9 become concentric circles centered at $(1/2, 3/2)$.

We get an "absolute" alternative to formula (8) by using $|x_0y_0|$ as the weight coordinate, instead of x_0y_0. This modification results in a distance function that is always non-negative. Its major drawback is that it does not distinguish between a point and its antipode. In particular, $\mathrm{dist}(p,p)$ and $\mathrm{dist}(p,\neg p)$ are both zero. Other alternatives based on the "absolute" vector addition formula (5, chapter 16) do not seem very useful.

4.3. Closeness and shortness

The jump in the value of $\mathrm{dist}(p,q)$ from $+\infty$ to $-\infty$ as p or q cross Ω may be a nuisance in geometric algorithms, since it means the numerical distance does not increase monotonically as the two points move apart. We can solve this problem by measuring the separation of two points by the reciprocal of formula (8), which I call the *closeness* of the two points:

$$\mathrm{cls}(x,y) = 1/\,\mathrm{dist}(x,y)$$

$$= \left[\, \sqrt{(y_0x_1 - x_0y_1)^2 + \cdots + (y_0x_\nu - x_0y_\nu)^2} \,,\; x_0y_0 \,\right] \qquad (9)$$

Similarly, we can define the *shortness* of a vector x of $\mathbf{V}_\nu$ as the reciprocal of its length, that is, the closeness of x and the front origin:

$$\mathrm{shr}(x) = 1/\,\mathrm{len}(x) = \left[\, \sqrt{(x_1)^2 + \cdots (x_\nu)^2} \,,\; x_0 \,\right]$$

These quantities are either infinite or points of the front range of $\mathbf{T}_1$. More importantly, they are better behaved than $\mathrm{dist}(p,q)$ and $\mathrm{len}(x)$. As y moves straight away from x towards $\neg x$, the value of $\mathrm{cls}(x,y)$ decreases monotonically from $+\infty$ to $-\infty$. Thus, closeness can be used to sort points by distance from a fixed reference point. The closeness is positive if x and y are on the same range, zero if one of them is at infinity, negative if they are on opposite ranges, and undefined (i.e., $\mathbf{0}$) if both arguments are at infinity. See figure 10.

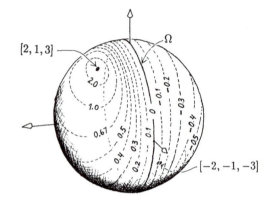

Figure 10. Closeness to $[2,1,3]$ on the two-sided plane.

The shortness of a vector x decreases monotonically from $+\infty$ to $-\infty$ as x moves straight away from O towards $\neg O$. It is positive for front vectors, zero for infinite vectors, and negative for back vectors.

4.4. Congruence and length ratio

It is important to note that length, shortness, distance, and closeness are *not* Euclidean concepts. The definitions given above do not carry over to arbitrary Euclidean spaces, since they assign special role to the origin O (as the vector with zero length) and implicitly define a unit of length, and these things are not preserved by arbitrary Euclidean maps.

What we *can* do in abstract Euclidean geometry is measure the *ratio* of two distances, and in particular check if they are congruent. Given four points $[a]$, $[b]$, $[x]$, and $[y]$ of $\mathbf{T}_\nu$, I define the *(two-sided) ratio of ab to xy* by the formula

$$\left[a_0 b_0 \sqrt{\sum_{k=1}^{\nu}(x_0 y_i - y_0 x_i)^2} \, , \, x_0 y_0 \sqrt{\sum_{k=1}^{\nu}(a_0 b_i - b_0 a_i)^2} \right] \tag{10}$$

This ratio is conserved by arbitrary Euclidean maps. It is a front positive point of $\mathbf{T}_1$ if and only if a lies on the same range as b and x lies on the same range as y.

Of course, if we all we need is to compare the length two segments, without computing those lengths, then we need only evaluate the squares of the two coordinates of formula (10), and compare them:

$$(a_0 b_0)^2 \left(\sum_{k=1}^{\nu}(x_0 y_i - y_0 x_i)^2 \right) \quad : \quad (x_0 y_0)^2 \left(\sum_{k=1}^{\nu}(a_0 b_i - b_0 a_i)^2 \right) \tag{11}$$

5. Angular measure and congruence

In classical Euclidean geometry we can measure not only lengths and distances, but angles as well. We can compare the angles between two pairs of lines by superimposing one on the other by means of a Euclidean transformation. In the same way we can compute the ratio between two angles, and the measure of an angle.

Extending these notions to two-sided geometry is relatively straightforward. In fact, we will see that in two-sided geometry angles can be handled somewhat more elegantly than in classical geometry, because they can be treated as points, and hence operated upon with the geometric tools we already have at our disposal.

5.1. Angles as real numbers

Observe that the angle between two proper lines in the straight model is only a function of their directions, and not of their absolute positions. Therefore, measuring the angle between two lines of $\mathbf{E}_\nu$ corresponds to measuring the separation between two points on Ω_ν.

Let $x = [0, x_1, x_2]$ be a point at infinity of $\mathbf{T}_2$. From Cartesian geometry, we know that the angle between the cardinal direction $\mathbf{e}^1 = [0, 1, 0]$ and x is

$$\arg(x) = \begin{cases} \text{arc } \tan(x_2/x_1) & \text{if } x_1 > 0 \\ \text{sign}(x_2) \cdot (\pi/2) & \text{if } x_1 = 0 \\ \text{arc } \tan(x_2/x_1) + \pi & \text{if } x_1 < 0 \end{cases} \qquad (12)$$

(Incidentally, this function is available in most programming languages and numerical libraries, usually as a two-argument arc-tangent procedure.)

Measuring angles by formula (12), or any other real-valued formula, is inconvenient for two reasons. First, it forces us to choose a unit of angular measure, even though angles are dimensionless quantities. Second, it forces us to choose one particular numeric value for the angle, even though the geometry only defines this value only up to a multiple of 2π radians. (As a consequence, real-valued angular measures are necessarily discontinuous.)

5.2. Angles as points at infinity

We can avoid both problems by representing angles as points of Ω_2, without reducing them to real numbers. To demonstrate this point, in what follows I will use the point $[0, x_1, x_2]$ to represent the angle between the vectors $(1, 0)$ and (x_1, x_2) of $\mathbf{R}^2$. In other words, I will represent the angle α by the point $[0, \cos\alpha, \sin\alpha]$. So, for example, $[0, 1, 0]$ is $0°$, $[0, 0, 1]$ is $90°$, $[0, -1, 0]$ is $180°$, $[0, 1, 1]$ is $45°$, and so on.

5.3. Angle arithmetic

This representation avoids the problems of real angular measures, but still lets us add, subtract, and compare angles using only the four arithmetic operations. Let us denote the operations on angles by $\alpha \hat{+} \beta$, $\alpha \hat{-} \beta$ and $\hat{-}\alpha$; their corresponding formulas are

$$[0, x_1, x_2] \hat{+} [0, y_1, y_2] = [0, x_1y_1 - x_2y_2, x_2y_1 + x_1y_2]$$

$$[0, x_1, x_2] \hat{-} [0, y_1, y_2] = [0, x_1y_1 + x_2y_2, x_2y_1 - x_1y_2] \qquad (13)$$

$$\hat{-}[0, x_1, x_2] = [0, x_1, -x_2]$$

We can understand these formulas by viewing the $[0, x_1, x_2]$ as the the complex number $x_1 + ix_2$, and the angle represented by the former as the argument of the latter. Then equations (13) can be recognized as the formulas for multiplication, division, and conjugation of complex numbers, except that the formula for $x \hat{-} y$ omits division of the result by the positive real number $\sqrt{x_0x_0 + x_1x_1}\sqrt{y_0y_0 + y_1y_1}$.

For example, the angle between two lines l and m of $\mathbf{T}_2$ is simply the angular difference between their directions, that is, $\mathrm{dir}(m) \hat{-} \mathrm{dir}(l)$. If $l = \langle l^0, l^1, l^2 \rangle$ and $m = \langle m^0, m^1, m^2 \rangle$, we have $\mathrm{dir}(l) = [0, l^2, -l^1]$ and $\mathrm{dir}(m) = [0, m^2, -m^1]$, and therefore

$$\mathrm{ang}(l, m) = [0, \ m^2l^2 + m^1l^1, \ m^2l^1 - m^1l^2] \qquad (14)$$

This formula can also be derived by computing the angular difference between normals, instead of directions. It is not hard to see that $\mathrm{ang}(l, m)$ is defined for any two proper lines of $\mathbf{E}_2$, and ranges over all points of Ω_2.

It can be shown that the result of formula (14) does not change if m and l are transformed by the same Euclidean map of $\mathbf{E}_\nu$ to itself. Therefore, the angle between two lines in an arbitrary two-dimensional Euclidean space can be measured by mapping the two lines into $\mathbf{E}_2$ with any Euclidean map φ, and applying formula (14) to the two images; the result will not depend on the map φ.

5.4. Angles in higher dimensions

In spaces with three or more dimensions, the angle between two lines l and m is even less well-defined than on the plane. The directions $\mathrm{dir}(l)$ and $\mathrm{dir}(m)$ — two points on the hyperplane Ω — are well defined, but the sign of the angle between them is not, because there is no sense in asking whether the line at infinity $\mathrm{dir}(l) \vee \mathrm{dir}(m)$ is oriented in the "positive" or "negative" sense. Therefore, we cannot distinguish an angle α from the angle $-\alpha$. Given two directions $x = [0, x_1, .. x_\nu]$

and $y = [0, y_1, \ldots y_\nu]$, the best we can do is compute the co-sine of the angle between them,

$$\cos \mathrm{ang}(x, y) = \frac{x_1 y_1 + \cdots + x_\nu y_\nu}{\sqrt{x_1 x_1 + \cdots + x_\nu x_\nu} \sqrt{y_1 y_1 + \cdots + y_\nu y_\nu}}$$

For consistency, we may want to represent this angle as the point $[0, c, \sqrt{1 - c^2}]$ of Ω_2, where $c = \cos \mathrm{ang}(x, y)$. Note that this angle is always between 0 and π radians, inclusive.

6. Non-Euclidean geometries

The topic of the last three chapters has been the use of oriented projective space to emulate Euclidean geometry and some related theories. To conclude this topic, let's consider briefly the modeling of two popular non-Euclidean geometries, *hyperbolic* and *elliptic* [1].

For hyperbolic geometry, we can use a two-sided version of Beltrami's model, consisting of all flats that intersect the unit ball of the front and back ranges of $\mathbf{T}_\nu$. That is, we take all points $[x_0, \ldots x_\nu]$ such that $x^1 x^1 + \cdots + x_\nu x_\nu \leq x_0 x_0$, and the flats obtained by joining those points. The result is a two-sided version of ν-dimensional hyperbolic space, with every point replaced by two antipodal points, every line replaced by two oppositely oriented lines, and so on. See figure 11.

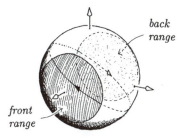

back range

front range

Figure 11. The two-sided hyperbolic plane.

The hyperbolic maps are then defined as the projective maps of $\mathbf{T}_\nu$ that take this set to itself. The notions of hyperbolic congruence and perpendicularity, and the hyperbolic metrics for distances and angles, can all be defined in terms of these maps.

For elliptic geometry, we can use the whole $\mathbf{T}_\nu$ as a model. The result is only a partial doubling of the standard ν-dimensional elliptic space: a single line of the standard model corresponds to two opposite lines of $\mathbf{T}_\nu$, but a point of the original corresponds to a single point of $\mathbf{T}_\nu$ (and vice-versa). The elliptic maps can be identified with the projective maps of $\mathbf{T}_\nu$ that preserve the standard polarity $\perp$.

As in the hyperbolic case, these maps determine completely the distance metric of elliptic geometry, which turns out to be just the great-circle distance of the spherical model. In fact, measuring angles in the Euclidean space $\mathbf{E}_\nu$ is equivalent to measuring distances on Ω_ν, treating the latter as the elliptic space of dimension $\nu - 1$. This equivalence is made evident by the map $[0, x_1, \ldots x_\nu] \mapsto [x_1, \ldots x_\nu]$, which takes Ω_ν to $\mathbf{T}_{\nu-1}$.

One advantage of these models is that many operations of hyperbolic and elliptic geometry (join, meet, map transformations, etc.) are then also operations of two-sided geometry, which means we do not have to write new code for them.

Chapter 18
Representing flats by simplices

In previous chapters we learned how to represent points and hyperplanes of $\mathbf{T}_\nu$ in a computer, using their homogeneous coordinates and coefficients, respectively. These encodings are sufficient for doing computational geometry on the plane, where points and hyperplanes (lines) are the only non-trivial flats. Problems in three or more dimensions, however, often involve flats of other ranks. In this chapter I will discuss a popular way of encoding such flats, the so-called *simplex representation*.

The simplex representation is an established tool in unoriented projective geometry. As we might expect, adapting it to the two-sided world does not require any additional arithmetic; it requires only that we pay a bit more attention to the signs of coordinates and to the order of operands.

1. The simplex representation

A flat of $\mathbf{T}_\nu$ generated by the κ-dimensional simplex a can be represented by the $k \times n$ real matrix

$$\begin{pmatrix} a_0^0 & a_1^0 & \dots & a_\nu^0 \\ \vdots & & & \vdots \\ a_0^\kappa & a_1^\kappa & \dots & a_\nu^\kappa \end{pmatrix} \tag{1}$$

whose rows are the homogeneous coordinates of the vertices of a.

As we saw in chapter 4, two κ-dimensional simplices are equivalent (determine the same flat) if and only if their matrices a and b satisfy $a = Kb$, where K is some $k \times k$ matrix with positive determinant. Therefore, the flats of $\mathbf{T}_\nu$ with rank k can be identified with the equivalence classes of the full-rank $k \times n$ matrices under this relation. (The matrices with less than full rank, i.e., with linearly dependent rows, can be identified with the null object $\mathbf{0}^k$.)

(As an aside, we can also use the simplex representation to represent unoriented flats; we have only to consider two coordinate matrices to be equivalent if they are related by a $k \times k$ matrix with *nonzero* determinant.)

I will write the matrix of a simplex a with square brackets to denote the class of all simplex matrices that are equivalent to it by the rule above. That is to say, if $a^0, \ldots a^\kappa$ is a proper simplex, with $a^i = [a_0^i, \ldots a_\nu^i]$, I will write

$$
a^0 \vee a^1 \vee \cdots \vee a^\kappa = \begin{bmatrix} a_0^0 & a_1^0 & \cdots & a_\nu^0 \\ \vdots & & & \vdots \\ a_0^\kappa & a_1^\kappa & \cdots & a_\nu^\kappa \end{bmatrix}
$$

By definition, a *coordinate matrix* for the flat $a^0 \vee \cdots \vee a^\kappa$ is any matrix equivalent to the one above.

In particular, the universe Υ of $\mathbf{T}_\nu$ is represented by the class of square $n \times n$ matrices with positive determinant, and its opposite $\neg \Upsilon$ by those with negative determinant. The vacua Λ and $\neg \Lambda$ must be treated as special cases, since the obvious representation (a $0 \times n$ matrix) doesn't distinguish between them.

1.1. Simplex orientation

Let $s = (s^0; \ldots s^\nu)$ be a ν-dimensional simplex of $\mathbf{T}_\nu$. The orientation of s is given by the sign of the determinant of its coordinate matrix. That is,

$$
\mathrm{sign}(s^0, \ldots s^\nu) = \mathrm{sign} \begin{vmatrix} s_0^0 & s_1^0 & \cdots & s_\nu^0 \\ s_0^1 & s_1^1 & \cdots & s_\nu^1 \\ \vdots & & & \vdots \\ s_0^\nu & s_1^\nu & \cdots & s_\nu^\nu \end{vmatrix}
$$

We can deduce many important properties of $\mathrm{sign}(s)$ directly from this definition. For instance, transposing any two vertices of a simplex reverses its sign, as does replacing any vertex by its antipode.

As another example, a cyclic permutation of all vertices preserves the sign of the simplex if the rank n is odd, and reverses it if n is even. This is because a cyclic permutation of n objects is equivalent to $n-1$ transpositions. For example, if $(p; q; r)$ is a positive triangle of $\mathbf{T}_2$, so are its cyclic permutations $(q; r; p)$ and $(r; p; q)$. On the other hand, if $(p; q; r; s)$ is a positive tetrahedron of $\mathbf{T}_3$, then $(q; r; s; p)$ and $(s; p; q; r)$ are negative tetrahedra, whereas $(r; s; p; q)$ is a positive one.

1.2. Join, meet, and relative orientation

To compute the join of two flats in the simplex representation, we need only stack the coordinate matrix of the first operand on top of that of second operand (provided that their ranks add to at most n).

Other operations can be much harder. For instance, to check whether two matrices s, t represent the same flat, we must check whether there exists a positive $k \times k$ real matrix A such that $s = At$. Also, to compute $a^\vdash$ or $a^\dashv$ we must find a basis for the orthogonal complement of the row space of the matrix of a. To compute $a \wedge b$ we must find a suitably oriented basis for the intersection of the row spaces of their matrices. Finally, to compute $a \diamond b$ we must compute the sign of the determinant of the $n \times n$ matrix that results from their join.

All these computations can be carried out in practice by Gaussian elimination, or any equivalent method, in roughly $O(nk^2)$ steps. When coding the algorithms, we must be careful to preserve the orientation implicit in the order of the rows. For example, when swapping two rows of a simplex we must also negate the elements of one row.

2. The dual simplex representation

Recall that an hyperplane h of $\mathbf{T}_\nu$ can be represented by n homogeneous coefficients h^j, such that $x \diamond h = \mathrm{sign}(x_0 h^0 + \cdots + x_\nu h^\nu)$ for any point $x = [x_0, .. x_\nu]$. Hyperplane coefficients are the basis of the *dual simplex representation* for flats of arbitrary rank. The idea is to represent a flat a of $\mathbf{T}_\nu$ with co-rank k by the coefficients of k hyperplanes $h_0, .. h_\kappa$ whose meet, in that order, is the flat a. It is convenient to write those numbers as an $n \times k$ *coefficient matrix*

$$
\begin{pmatrix}
a_0^0 & \cdots & a_\kappa^0 \\
a_0^1 & & a_\kappa^1 \\
\vdots & & \vdots \\
a_0^\nu & \cdots & a_\kappa^\nu
\end{pmatrix}
\tag{2}
$$

where column j gives the coefficients of the jth hyperplane.

Observe that in this case we have $a^\dashv = (a_0)^\dashv \vee \cdots \vee (a_\kappa)^\dashv$. The homogeneous coordinates of the point $(a_i)^\dashv$ are the coefficients of the hyperplane a_i, viewed as a row vector: $(a_i)^\dashv = [a_i^0, .. a_i^\nu]$. It follows that the (primal) simplex representation of the flat $a^\dashv$ is the $k \times n$ matrix whose rows are the columns of (2), in the same order. We conclude that *a coefficient matrix for a flat a is the transpose of a coordinate matrix for the flat $a^\dashv$.*

From this observation it follows that two coefficient matrices a, b denote the same flat of $\mathcal{F}_n^{n-k}$ if and only if there is a $k \times k$ matrix K with positive determinant such that $b = aK$. I will denote the class of coefficient matrices equivalent to (2) by horizontal square brackets above and below the matrix. That is, I will write

$$a_0 \wedge a_1 \wedge \cdots \wedge a_\kappa \;=\; \begin{array}{|ccc|} a_0^0 & \cdots & a_\kappa^0 \\ a_0^1 & & a_\kappa^1 \\ \vdots & & \vdots \\ a_0^\nu & \cdots & a_\kappa^\nu \end{array}$$

where $a_i = \langle a_i^0, \ldots a_i^\nu \rangle$. As expected from duality, the positive vacuum of $\mathbf{T}_\nu$ is represented by the class of $n \times n$ coefficient matrices with positive determinant, and the negative vacuum by those with negative determinant. The universes $+\Upsilon$ and $-\Upsilon$ must be handled as a special case, since their coefficient matrices have zero columns.

Note that every point of a flat is on every hyperplane containing that flat. We conclude that if A is the $k \times n$ coordinate matrix of a flat, and B is its $n \times (n - k)$ coefficient matrix, then the product AB is the zero matrix of size $k \times (n - k)$. Conversely, if this is true of two full-rank matrices A and B, then they denote the same flat, except for orientation.

Symmetrically, the dual simplex representation makes it quite easy to compute the meet of two flats: it suffices to glue their coefficient matrices side by side. On the other hand, other operations (including join) require some variant of the Gaussian elimination algorithm.

2.1. Mixed representation

The simplex representation is highly redundant, especially when k is close to n. The condition for equivalence stated above implies that the set of all flats of rank k in $\mathbf{T}_\nu$ has dimension $kn - k^2$, and yet their coordinate matrices have kn coefficients. For example, the matrix of a hyperplane has an $n \times (n - 1)$ elements, when n would be more than enough.

We can alleviate this problem by using a *mixed representation*, in which a flat is represented by either a simplex or a dual simplex, whichever is smaller. That is, we use the coordinate matrix if the rank k is $< n/2$, and the coefficient matrix when $k > n/2$ (when $k = n/2$ we can use either form). This mixed strategy reduces the wasted storage, but doesn't eliminate it completely. The maximum waste is now 50%, which occurs occurs with flats of intermediate rank $k \approx n/2$: they use $\approx n^2/2$ elements but have only $\approx n^2/4$ degrees of freedom.

One advantage of this mixed representation is that the polar complement operations $\vdash$ and $\dashv$ then become trivial.

2.2. Converting between coordinates and coefficients

It is often necessary to compute the dual matrix of a flat from the primal one, or vice-versa. In particular, if we use the mixed representation described above, we will often have to do this conversion as part of join or meet operations. As discussed above, this problem can be reduced to that of computing the coordinate matrix of $a^\vdash$ given that of a, which can be done by Gaussian elimination.

3. The reduced simplex representation

A more promising way to reduce storage costs and the ambiguity of the matrix representation is to represent each flat by some "canonical" simplex. If we choose canonical simplices with matrices of a particular simple form, we can usually encode the latter with far less than kn elements.

In particular, we may consider the *reduced simplices* whose coordinate matrices have the form described below. I will say that a $k \times n$ matrix a is *reduced* if there are k integers $0 \le j_0 < j_1 < \cdots < j_\kappa \le \nu$ (the *pivot columns*) such that, for all i,

(i) $a^i_{j_i}$ is ± 1 if $i = 0$, and 1 if $i > 0$.

(ii) $a^i_{j_i}$ is the only nonzero element in column j_i,

(iii) $a^i_{j_i}$ is the first nonzero element in row i, for all i.

For example, here is a reduced simplex of rank 4 in $\mathbf{T}_9$:

$$
\begin{array}{cccccccccc}
 & j_0 & & & j_1 & & j_2 & j_3 & & \\
 & \downarrow & & & \downarrow & & \downarrow & \downarrow & & \\
\left(\begin{array}{cccccccccc}
0 & 1 & \boxed{3} & \boxed{-6} & 0 & \boxed{4} & 0 & 0 & \boxed{5} & \boxed{-1} \\
0 & 0 & 0 & 0 & 1 & \boxed{3} & 0 & 0 & \boxed{3} & \boxed{7} \\
0 & 0 & 0 & 0 & 0 & 0 & 1 & 0 & \boxed{-6} & \boxed{0} \\
0 & 0 & 0 & 0 & 0 & 0 & 0 & 1 & \boxed{9} & \boxed{1}
\end{array}\right)
\end{array}
\tag{3}
$$

Conditions (i) and (ii) say that the $k \times k$ sub-matrix of a formed by columns $j_0, \ldots j_\kappa$ is the identity matrix, except that its first element may be -1. Condition (iii) implies that any other matrix equivalent to a that satisfies (i) and (ii) will have a sequence of pivot columns that is lexicographically larger than $j_0, \ldots j_\kappa$. It follows that there is at most one reduced matrix equivalent to a given one. In other words, every flat contains at most one reduced simplex.

On the other hand, we can convert any simplex to an equivalent reduced matrix by a straightforward variant of the Gaussian elimination algorithm. The only novelty is that during the algorithm we must watch for operations that reverse the orientation of the simplex (namely, swapping of two rows and multiplication of a row by a negative number), and compensate for them by also negating the first row.

This shows we can represent an arbitrary flat of $\mathbf{T}_\nu$ with rank k by a triplet (σ, J, R), where σ is a single bit that specifies the value of $a^0_{j_0}$ ($+1$ or -1), J is the set of pivot columns, and R is a linear array containing the remaining variable elements of the reduced simplex, namely a^i_j for $j > i, j \notin J$. These elements are shown boxed in example (3). This information occupies at most $k(n - k)$ floating-point words plus $n + 1$ bits of storage. (Again, we are not counting the space needed to store the ranks k and n.)

The operations of meet, join, and relative complement can be computed by expanding the the operands to the unreduced matrix form, computing an unreduced simplex for the result, and reducing it as described above. It is of course possible to combine all those steps into optimized algorithms that operate directly on the reduced operands and construct the result directly in reduced form. However, however, the relatively modest savings of memory space and arithmetic operations that would result must be balanced against the costs of increased program complexity and indexing overhead.

Chapter 19
Plücker coordinates

Plücker (or *Grassmann*) *coordinates* are a generalization of homogeneous coordinates to flats of arbitrary rank. The Plücker representation is mathematically more elegant than the simplex representation described in chapter 18, but is substantially more expensive — exponentially so — in time and space. Therefore, Plücker coordinates are valuable mostly in theoretical work, and in computations restricted to spaces of dimension four or less.

1.1. The Plücker coordinates of a line

According to theorem 1 of chapter 7, we can completely determine a flat a if we know its orientation relative to every flat of complementary rank. For instance, a line l of $\mathbf{T}_3$ is uniquely determined by the values of $l \diamond h$ when h ranges over all lines of $\mathbf{T}_3$. Let $(u; v)$ and $(x; y)$ be simplices spanning l and h, respectively. The value of $l \diamond h$ is then given by

$$l \diamond h = \text{sign} \begin{vmatrix} u_0 & u_1 & u_2 & u_3 \\ v_0 & v_1 & v_2 & v_3 \\ x_0 & x_1 & x_2 & x_3 \\ y_0 & y_1 & y_2 & y_3 \end{vmatrix}$$

We can expand this determinant into a sum of six terms, each being the product of a 2×2 minor from the first two rows, and the "complementary" minor from the bottom two rows. That is, $l \diamond h$ is the sign of

$$l_{\{01\}} h_{\{23\}} - l_{\{02\}} h_{\{13\}} + l_{\{12\}} h_{\{03\}} + l_{\{03\}} h_{\{12\}} - l_{\{13\}} h_{\{02\}} + l_{\{23\}} h_{\{01\}}.$$

where

$$l_{\{ij\}} = \begin{vmatrix} u_i & u_j \\ v_i & v_j \end{vmatrix} \quad \text{and} \quad h_{\{ij\}} = \begin{vmatrix} x_i & x_j \\ y_i & y_j \end{vmatrix}$$

It follows that the line l is uniquely determined by the six minors $l_{\{01\}}, l_{\{02\}}, \ldots, l_{\{23\}}$. These six numbers are the *Plücker coordinates* of the line.

Observe that the choice of the simplex $(u; v)$ affects the Plücker coefficients of l only by a positive factor. That is, if $(p; q) = A(u; v)$ for some 2×2 matrix A with positive determinant, then the Plücker coefficients $m_{\{ij\}}$ computed from $(p; q)$ will satisfy $m_{\{ij\}} = |A| \, l_{\{ij\}}$. For example, the line l determined by the simplex

$$\begin{pmatrix} 1 & 2 & 0 & 3 \\ 0 & 2 & 1 & 0 \end{pmatrix} \tag{1}$$

has Plücker coordinates

$$l_{\{01\}} = \begin{vmatrix} 1 & 2 \\ 0 & 2 \end{vmatrix} = 2 \qquad l_{\{02\}} = \begin{vmatrix} 1 & 0 \\ 0 & 1 \end{vmatrix} = 1 \qquad l_{\{12\}} = \begin{vmatrix} 2 & 0 \\ 2 & 1 \end{vmatrix} = 2$$

$$l_{\{03\}} = \begin{vmatrix} 1 & 3 \\ 0 & 0 \end{vmatrix} = 0 \qquad l_{\{13\}} = \begin{vmatrix} 2 & 3 \\ 2 & 0 \end{vmatrix} = -6 \qquad l_{\{23\}} = \begin{vmatrix} 0 & 3 \\ 1 & 0 \end{vmatrix} = -3$$

Few computer graphics programmers seem to be aware of this six-number encoding of lines in three-space. (One notable exception is Patrick Hanrahan's geometry calculator [12].) A line is usually represented as a pair (point, direction vector), as a pair of points, or as the intersection of two planes. In three dimensions, all those encodings are generally less elegant and less efficient than the Plücker coordinate representation.

1.2. Plücker coordinates for general flats

The analysis that led us to the Plücker coordinates of a line is equally valid for a flat a of arbitrary rank k in any space $\mathbf{T}_\nu$. The flat a is uniquely determined by the values of $a \diamond h$ where h ranges over all flats of $\mathbf{T}_\nu$ with rank $m = n - k$. Therefore, if s is any positive simplex of a, and x is a positive simplex of h, then

$$a \diamond h = \mathrm{sign} \begin{vmatrix} s_0^0 & s_1^0 & \cdots & \cdots & s_\nu^0 \\ \vdots & \vdots & & & \vdots \\ s_0^\kappa & s_1^\kappa & \cdots & \cdots & s_\nu^\kappa \\ x_0^0 & x_1^0 & \cdots & \cdots & x_\nu^0 \\ \vdots & \vdots & & & \vdots \\ x_0^\mu & x_1^\mu & \cdots & \cdots & x_\nu^\mu \end{vmatrix} \tag{2}$$

We can expand the determinant (2) into a sum of terms, each being the product of a $k \times k$ minor determinant from the first k rows, and an $m \times m$ minor determinant from the last m rows:

$$a \diamond h = \sum_{\substack{I \cap J = \emptyset \\ I \cup J = \{0..\nu\} \\ |I|=k, |J|=m}} (-1)^{|I>J|} a_I h_J \tag{3}$$

where

$$a_{\{i_0, i_1, ..., i_\kappa\}} = \begin{vmatrix} s^0_{i_0} & \cdots & s^0_{i_\kappa} \\ \vdots & & \vdots \\ s^\kappa_{i_0} & \cdots & s^\kappa_{i_\kappa} \end{vmatrix} \tag{4}$$

and

$$h_{\{j_0, j_1, ... j_\mu\}} = \begin{vmatrix} x^0_{j_0} & \cdots & x^0_{j_\mu} \\ \vdots & & \vdots \\ x^\mu_{j_0} & \cdots & x^\mu_{j_\mu} \end{vmatrix} \tag{5}$$

The exponent $|I > J|$ in formula (3) is the number of pairs i, j with $i \in I$, $j \in J$, and $i > j$. In other words, $|I > J|$ is the number of inversions in the permutation of $\{0..\nu\}$ that consists of all elements of I followed by those of J.

Formula (3) says that the flat a is uniquely determined by the $\binom{n}{k}$ minor determinants a_I. Those numbers are by definition the *Plücker* or *homogeneous coordinates* of a. The set $\{i_0, i_1, \ldots, i_\kappa\}$ of the columns included in minor (4) is the *label* of that coordinate. The labels range over all k-element subsets of $N = \{0, ..\nu\}$.

The Plücker coordinates of a flat cannot be all zero. As we know from linear algebra, the determinants (4) are all zero if and only if the rows of the matrix

$$\begin{pmatrix} s^0_0 & s^0_1 & \cdots & s^0_\nu \\ \vdots & & & \vdots \\ s^0_0 & s^0_1 & \cdots & s^0_\nu \end{pmatrix}$$

are linearly dependent, that is, if the simplex s is degenerate.

As in the case of lines of $\mathbf{T}_3$, the particular choice of the representative simplex s (and any scaling of the homogeneous coordinates of each s^i) will affect the Plücker coordinates of a only to the extent of multiplying all of them by some positive real number. We conclude that *two flats of* $\mathbf{T}_\nu$ *are the same flat if and only if their Plücker coordinates differ only by a positive factor.*

Because of this scale ambiguity, it doesn't make sense to refer to the magnitude of a Plücker coordinate in isolation, but only to a complete set of them. We might consider normalizing the coordinates in some way (e.g., so that their squares add to 1). However, as in the case of point coordinates, this normalization is usually not worth its cost, except perhaps for the purpose of avoiding arithmetic overflow.

1.3. The natural order of Plücker coordinates

In formulas and in computer programs it is more convenient to write the Plücker coordinates of a flat in some canonical order, so that the labels can be omitted. A convenient choice is to enumerate the labels in increasing order of their *binary value.*

By definition, the binary value of a finite set X of natural numbers is $\text{binv}\, X = \sum_{x \in X} 2^x$. Note that if we write $\text{binv}\, X$ in base two, then the elements of X are the positions of the "1" bits, from right to left. For example,

$$\begin{matrix} 6 & 43 & & 0 \\ \downarrow & \downarrow\downarrow & & \downarrow \end{matrix}$$
$$\text{binv}\, \{0\,3\,4\,6\} = 1011001_2 = 89_{10}$$

When dealing with Plücker coordinates we often need to consider the collection of all subsets of $\mathbf{N}$ with a fixed size k. I will denote by $k{:}i$ the ith subset of this list, in order of increasing binary value, starting from $i = 0$. If $k{:}i = X$, I will call i the *index* of X, and write $i = \#X$. Note that $\#X$ is usually different from (and much smaller than) $\text{binv}\, X$. For example, the the first few three-element subsets of $\mathbf{N}$ and their binary values are

3:0	3:1	3:2	3:3	3:4	3:5	3:6	3:7	3:8	3:9	3:10	⋯
{0 1 2}	{0 1 3}	{0 2 3}	{1 2 3}	{0 1 4}	{0 2 4}	{1 2 4}	{0 3 4}	{1 3 4}	{2 3 4}	{0 1 5}	⋯
7	11	13	14	19	21	22	25	26	28	35	⋯

In general, a set I precedes another set J in this ordering if $\max I < \max J$, or $\max I = \max J$ and $I \setminus \{\max I\}$ precedes $J \setminus \{\max J\}$. We can obtain this ordering also by writing each subset as a *decreasing* sequence of numbers, and sorting those sequences in *increasing* lexicographic order. Note that for any j all subsets of $\{0 .. j\}$ occur together at the beginning of the list, and before any subset involving elements greater than j.

I will use the notation $[z_0, z_1, \ldots]^k$ for the flat of $\mathbf{T}_\nu$ with rank k whose Plücker coordinates are $z_0, z_1, \ldots$, listed in the natural order of their labels. I.e., the label set of z_i is $k{:}i$. For example, I will denote the line of example (1) as

$$l = [\ 2, \quad 1, \quad 2, \quad 0, \quad -6, \quad -3\ ]^2$$
$$\begin{matrix} \uparrow & \uparrow & \uparrow & \uparrow & \uparrow & \uparrow \\ \{0\,1\} & \{0\,2\} & \{1\,2\} & \{0\,3\} & \{1\,3\} & \{2\,3\} \end{matrix}$$

With this convention, the uniqueness and equivalence properties of Plücker coordinates can be restated as follows: *For any k, $[x_0, x_1, \ldots]^k$ and $[y_0, y_1, \ldots]^k$ are the same flat if and only if $x_i = \alpha y_i$ for some $\alpha > 0$ and all i.*

1.4. Points and hyperplanes

Note that for points ($k = 1$) the compact Plücker notation defined above coincides with the standard homogeneous notation. That is, the point with homogeneous coordinates $[u_0, u_1, \ldots]$ has Plücker coordinates $[u_0, u_1, \ldots]^1$.

A hyperplane ($k = n - 1$) has n Plücker coordinates, whose labels are the sets $\{0, \ldots \nu\} \setminus i$ for each i. These coordinates are the coefficients of the hyperplane as commonly used in graphics programming, except that they are listed in reverse order and half the signs are reversed. The details will be given in chapter 20.

1.5. Lines on the plane

A line l of $\mathbf{T}_2$ has three Plücker coordinates, namely $l_{\{01\}}, l_{\{02\}}, l_{\{12\}}$. If $(u; v)$ is a positive simplex of l with $u = [u_0, u_1, u_2]$ and $v = [v_0, v_1, v_2]$, then

$$
l_{\{01\}} = \begin{vmatrix} u_0 & u_1 \\ v_0 & v_1 \end{vmatrix} \qquad l_{\{02\}} = \begin{vmatrix} u_0 & u_2 \\ v_0 & v_2 \end{vmatrix} \qquad l_{\{12\}} = \begin{vmatrix} u_1 & u_2 \\ v_1 & v_2 \end{vmatrix}
$$

Here are some lines of $\mathbf{T}_2$, and their Plücker coordinates:

$$
\begin{aligned}
\mathbf{e}^0 \vee \mathbf{e}^1 &= [1, 0, 0]^2 & &\text{the } x\text{-axis} \\
\mathbf{e}^0 \vee \mathbf{e}^2 &= [0, 1, 0]^2 & &\text{the } y\text{-axis} \\
\mathbf{e}^1 \vee \mathbf{e}^2 &= [0, 0, 1]^2 & &\Omega_2 \\
\mathbf{e}^0 \vee [1, 1, 1] &= [1, 1, 0]^2 & &\text{bisector of first quadrant} \\
[1, 1, 0] \vee [1, 0, 2] &= [-1, 2, 2]^2 & &\text{see figure 1.}
\end{aligned}
$$

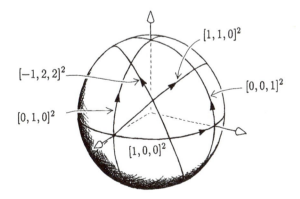

Figure 1. Some lines of $\mathbf{T}_2$ and their Plücker coordinates.

1.6. Canonical flats

Let $I = \{i_0, i_1, \ldots, i_\kappa\}$ be any subset of $N = \{0, \ldots \nu\}$ with size k and $i_0 < i_1 < \cdots < i_\kappa$. By definition, the join of the canonical points $e^{i_0}, \ldots e^{i_\kappa}$, in that order, is the *canonical flat of* T_ν *with label* I, denoted by e^I. Its Plücker coordinates are all zero, except for $(e^I)_I = 1$.

For example, the canonical lines of T_3 are

$$e^{\{0\,1\}} = e^0 \vee e^1 = [1\,0\,0\,0\,0\,0]^2 \quad x\text{-axis}$$

$$e^{\{0\,2\}} = e^0 \vee e^2 = [0\,1\,0\,0\,0\,0]^2 \quad y\text{-axis}$$

$$e^{\{1\,2\}} = e^1 \vee e^2 = [0\,0\,1\,0\,0\,0]^2 \quad \text{line at infinity on the } xy \text{ plane}$$

$$e^{\{0\,3\}} = e^0 \vee e^3 = [0\,0\,0\,1\,0\,0]^2 \quad z\text{-axis}$$

$$e^{\{1\,3\}} = e^1 \vee e^3 = [0\,0\,0\,0\,1\,0]^2 \quad \text{line at infinity on the } xz \text{ plane}$$

$$e^{\{2\,3\}} = e^2 \vee e^3 = [0\,0\,0\,0\,0\,1]^2 \quad \text{line at infinity on the } yz \text{ plane.}$$

1.7. Vacuum and universe

The universe of T_ν and its antipode have a single coordinate, whose label is the entire set $\{0 \ldots \nu\}$. It is easy to see that $\Upsilon_\nu = [+1]^n$, and $\neg\Upsilon_\nu = [-1]^n$. By convention, the flats of zero rank (the vacua) also have a single coordinate, whose label is the empty set and whose sign describes the flat's orientation; that is, $\Lambda = [+1]^0$ and $\neg\Lambda = [-1]^0$.

2. The canonical embedding

Recall that the canonical embedding of T_ν into a space T_μ with $\mu > \nu$ is obtained by appending $\mu - \nu$ zeros to the homogeneous coordinates of every point. In general, if a is a flat of rank k in T_ν, its canonical embedding $\hat{a}$ is obtained by appending $\mu - \nu$ zero columns at the right end of its coordinate matrix:

$$\begin{pmatrix} s_0^0 & s_1^0 & \cdots & s_\nu^0 \\ \vdots & & & \vdots \\ s_0^0 & s_1^0 & \cdots & s_\nu^0 \end{pmatrix} \mapsto \begin{pmatrix} s_0^0 & s_1^0 & \cdots & s_\nu^0 & 0 & \cdots & 0 \\ \vdots & & & \vdots & \vdots & & \vdots \\ s_0^0 & s_1^0 & \cdots & s_\nu^0 & 0 & \cdots & 0 \end{pmatrix}$$

What is the effect of this embedding in terms of Plücker coordinates? The flat $\hat{a}$ has $\binom{m}{k}$ coordinates, each labeled with a subset of $\{0 \ldots \mu\}$ with size k. Now observe that in the natural order of these sets, all those that are subsets of $\{0 \ldots \nu\}$ occur together at the beginning of the list. Therefore, the first $\binom{n}{k}$ Plücker coordinates of $\hat{a}$ are exactly the coordinates of a. Moreover, any other coordinate of $\hat{a}$ is zero, since it is the determinant of a $k \times k$ matrix that includes at least one zero column. We conclude that *the canonical embedding of* $\mathbf{T}_\nu$ *into* $\mathbf{T}_\mu$ *merely appends* $\binom{m}{k} - \binom{n}{k}$ *zeros to the Plücker coordinates of every flat of rank* k.

3. Plücker coefficients

The *Plücker coefficients* of general flats are related to the homogeneous coefficient of hyperplanes in the same way that Plücker coordinates are related to the homogeneous coordinates of points. If a is a flat with coefficient matrix

$$
\begin{bmatrix}
h_0^0 & \cdots & h_\kappa^0 \\
h_0^1 & & h_\kappa^1 \\
\vdots & & \vdots \\
h_0^\nu & \cdots & h_\kappa^\nu
\end{bmatrix}
$$

(that is, $a = h_0 \wedge \cdots \wedge h_\kappa$, where $h_j = \langle h_j^0, .. \rangle$), then the Plücker coefficients of a are by definition the $k \times k$ minor determinants of that matrix:

$$
a^{\{j_0, j_1, \ldots, j_\kappa\}} = \begin{vmatrix}
h_0^{j_0} & \cdots & h_0^{j_\kappa} \\
\vdots & & \vdots \\
k_\kappa^{j_0} & \cdots & h_\kappa^{j_\kappa}
\end{vmatrix}
$$

As in the case of coordinates, it is convenient to list the Plücker coefficients of a flat in the natural order of their label sets. I will denote the ith element of this list (counting from 0) by a^i; i is the *index* of the coefficient. I will also write $\langle c^0, c^1, \ldots \rangle^k$ to denote the flat of co-rank k with Plücker coefficients $a^0, a^1, \ldots$. If the flat a is an hyperplane, its Plücker coefficients coincide with its homogeneous coefficients are the same; that is, $\langle a^0, a^1, \ldots \rangle^1 = \langle a^0, a^1, \ldots \rangle$. Like the Plücker coordinates, the Plücker coefficients are unique only up to a positive scaling factor.

The Plücker coefficients of a flat a are distinct from but closely related to the Plücker coordinates of a. As we shall see in the next chapter, to convert from one representation to the other we have to reverse the order of all elements, and negate some of them. In particular, a line of $\mathbf{T}_2$ with coefficients $\langle a, b, c \rangle$ has Plücker coordinates $[c, -b, a]^2$, and conversely.

4. Storage efficiency

The Plücker representation for a flat of $\mathbf{T}_\nu$ with rank k requires $\binom{n}{k}$ coordinates, versus the kn required by the simplex representation. Obviously, Plücker coordinates are far too expensive for large values of k and n. For $\nu \leq 4$, however, they are no more expensive than the simplex form, as shown below:

rank of space $= n$	2	3		4			5			
rank of flat $= k$	1	1	2	1	2	3	1	2	3	4
Simplex $= \min\{kn, (n-k)n\}$	2	3	3	4	8	4	5	10	10	5
Plücker $= \binom{n}{k}$	2	3	3	4	6	4	5	10	10	5
Reduced simplex $= k(n-k)+1$ [†]	2	3	3	4	5	4	5	7	7	5
$\dim \mathcal{F}_n^k = k(n-k)$	1	2	2	3	4	3	4	6	6	4

(6)

[†] Assuming that the sign bit and the pivot indices together use no more space than one matrix element.

As the table shows, for two-, three-, and four-dimensional geometry the Plücker coordinate representation is no bigger than the (unreduced) simplex representation. In fact, it is slightly smaller for lines in three-space (six numbers instead of eight). The reduced simplex form is somewhat more economical than the Plücker one, but it is not clear whether that is enough to offset its other drawbacks.

5. The Grassmann manifolds

According to the vector space model, the set $\mathcal{F}_n^k$ of all rank k flats of $\mathbf{T}_\nu$ is also the set of all k-dimensional oriented linear subspaces of $\mathbf{R}^n$. This set is the *oriented Grassmann manifold* [14, VII, XIV].

From the simplex representation, we know that every element of $\mathcal{F}_n^k$ is an equivalence class of $\mathbf{R}^{k \times n}$ (the $k \times n$ matrices), where two matrices are equivalent if one is obtained from the other through multiplication by an $k \times k$ matrix with positive determinant. The set of these matrices has dimension k^2. From these observations it follows eventually that $\mathcal{F}_n^k$ is a manifold of dimension $kn - k^2 = k(n-k)$. These numbers are listed as the bottom row of table (6).

Observe that in general there is a wide gap between the dimension of $\mathcal{F}_n^k$, on one hand, and the number of coordinates used by the simplex and Plücker representations, on the other. This gap is already evident for lines in three-space ($k = 2$, $n = 4$): the set of all such lines is only a four-dimensional manifold, but each line has six Plücker coordinates. One of these six degrees of freedom is "wasted" in the arbitrary scaling factor. The other is lost because not all sextuples $l_0, .. l_5$ of real numbers are the Plücker coordinates of some line. In fact, this happens if and only if the numbers satisfy the equation

$$l_0 l_5 - l_1 l_4 + l_2 l_3 = 0 \tag{7}$$

In general, a list of $\binom{n}{k}$ real numbers $z_0, z_1, \ldots$ can be interpreted as the Plücker coordinates of a flat of rank k of $\mathbf{T}_\nu$ if and only if they are not all zeros, and they satisfy a number of equations of the form

$$\sum_{i,j} \lambda_{ijr} z_i z_j = 0$$

for $r = 0, 1, \ldots$, where the coefficients λ_{ijr} are in $\{-1, 0, +1\}$. More details can be found in the book by Hodge and Pedoe [14, ch. VII].

Chapter 20
Formulas for Plücker coordinates

Let's now consider the problem of performing the basic operations of oriented projective geometry — join, meet, relative orientation, and polar complement — in terms of Plücker coordinates.

1. Algebraic formulas

Let's first derive general algebraic formulas that valid for all dimensions. It turns out that, for all the basic operations, the coordinates of the result are given by simple bilinear formulas: finite sums, where each term is a product of a sign coefficient ($+1$ or -1) and two Plücker coordinates, one from each operand.

1.1. Join

According to the definition, the join of m points $[u^0] \vee [u^1] \vee \cdots \vee [u^\mu]$ is the flat whose Plücker coordinates are all the $m \times m$ minor determinants of the matrix

$$\begin{pmatrix} u_0^0 & u_1^0 & \cdots & \cdots & u_\nu^0 \\ \vdots & & & & \vdots \\ u_0^\mu & u_1^\mu & \cdots & \cdots & u_\nu^\mu \end{pmatrix}$$

Let us now compute the join of two arbitrary flats a and b of $\mathbf{T}_\nu$, given their Plücker coordinates. Let $r = \operatorname{rank}(a)$, $s = \operatorname{rank}(b)$, $t = r + s = \operatorname{rank}(a \vee b)$ (we must of course have $t \leq n$ for the join to be defined). Let u and v be positive simplices of a and b, respectively; then $(u^0; \ldots u^\kappa; v^0; \ldots v^\mu)$ is a positive simplex of $a \vee b$. The Plücker coordinates of $c = a \vee b$ are therefore the $t \times t$ minor determinants of the $t \times n$ matrix

$$C = \begin{pmatrix} u_0^0 & u_1^0 & \cdots & \cdots & u_\nu^0 \\ \vdots & & & & \vdots \\ u_0^\rho & u_1^\rho & \cdots & \cdots & u_\nu^\rho \\ v_0^0 & v_1^0 & \cdots & \cdots & v_\nu^0 \\ \vdots & & & & \vdots \\ v_0^\sigma & v_1^\sigma & \cdots & \cdots & v_\nu^\sigma \end{pmatrix} \tag{1}$$

That is, the coordinates are the numbers

$$(a \vee b)_{\{k_0,..k_T\}} = c_{\{k_0,..k_T\}} = \begin{vmatrix} u_{k_0}^0 & u_{k_1}^0 & \cdots & \cdots & u_{k_T}^0 \\ \vdots & & & & \vdots \\ u_{k_0}^\rho & u_{k_1}^\rho & \cdots & \cdots & u_{k_T}^\rho \\ v_{k_0}^0 & v_{k_1}^0 & \cdots & \cdots & v_{k_T}^0 \\ \vdots & & & & \vdots \\ v_{k_0}^\sigma & v_{k_1}^\sigma & \cdots & \cdots & v_{k_T}^\sigma \end{vmatrix}$$

where $\{k_0,.. k_T\}$ ranges over all sorted subsets of $\{0,.. \nu\}$ of size t.

We can expand each determinant c_K in terms of the $r \times r$ minors of the first r rows, and the $s \times s$ minors of the last s rows, according to the formula

$$(a \vee b)_K = \sum_{\substack{I \cup J = K \\ I \cap J = \emptyset \\ |I|=r, |J|=s}} (-1)^{|I>J|} a_I b_J \qquad (2)$$

This formula is easily derived from the definition of determinants. Observe that the Plücker coordinates c_K of $a \vee b$ are indeed bilinear functions of the coordinates a_I, b_J of a and b.

1.2. Incidence

The formulas for join also give us a way to test whether a point lies on a flat, or whether two flats intersect. Recall that a point x is incident to a flat a if and only if $a \vee x = \mathbf{0}$ (provided we consider x and $-x$ to be incident to each other, too). In general, flats a and b intersect if and only if $a \vee b = \mathbf{0}$.

Algebraically, two flats a and b of $\mathbf{T}_\nu$ have a point in common if and only if the rows of the join matrix C in (1) are not linearly independent. In this case, the minor determinants c_K are all zero. This suggests we define the Plücker coordinates of the null object $\mathbf{0}^k$ as being $[0, 0, \ldots, 0]^k$. With this convention, formula (2) will automatically return the correct value in all cases. In fact, the test $a \vee b = \mathbf{0}$ is a simple way to check whether two flats a and b intersect, and in particular whether a point a lies on a flat b. In general, for two flats of ranks r and s in $\mathbf{T}_\nu$, this method tells us to evaluate $\binom{n}{r+s}$ bilinear functions of the coordinates, and check whether they are all zero.

In particular, a point $x = [x_0, \ldots x_\nu]$ lies on a flat a if and only if $x \vee a = \mathbf{0}$, i.e.

$$\sum_{i \in K} (-1)^{|i > K \setminus i|} x_i\, a_{K \setminus i} = 0 \qquad (3)$$

for all $K \subseteq N$ with $|K| = \text{rank}(A) + 1$. For example, a point $x = [x_0, \ldots x_3]$ of $\mathbf{T}_3$ is on the line $a = [a_0, \ldots a_5]^2$ if and only if

$$\begin{cases} a_2 x_0 - a_1 x_1 + a_0 x_2 \qquad\quad = 0 \\ a_4 x_0 - a_3 x_1 \qquad\quad + a_0 x_3 = 0 \\ a_5 x_0 \qquad\quad - a_3 x_2 + a_1 x_3 = 0 \\ \qquad\quad a_5 x_1 - a_4 x_2 + a_2 x_3 = 0 \end{cases} \qquad (4)$$

1.3. Independent hyperplanes defining a flat

Formulas (4) above are rather inefficient, since testing all of them requires 12 multiplications, 8 additions, and 4 tests for zero. For comparison, if we represented the line a as the intersection of two independent planes, we could test for incidence with 8 multiplications, 6 additions, and 2 tests for zero.

However, note that the four conditions (4), viewed as linear equations on the variables x_i with coefficients a_j, cannot be all linearly independent. Since they are necessary and sufficient conditions for x to be on the line a, and a is a flat of rank 2, their solutions must span a two-dimensional linear space. So, we must be able to write two of those equations as linear combinations of the other two, which must be linearly independent. These two independent equations determine two distinct unoriented planes whose intersection is the line a.

Therefore, a more efficient way of testing a point against a given line is to first look for two linearly independent equations in system (4), and then evaluate only those two equations. To find such equations, we need only find a non-zero line coordinate a_j, and pick the two equations where that coordinate appears as a coefficient. By inspection one can check that the two equations have the form

$$\begin{aligned} \cdots \pm a_j x_i \pm \cdots + 0 \cdot x_k \pm \cdots = 0 \\ \cdots + 0 \cdot x_i \pm \cdots \pm a_j x_k \pm \cdots = 0 \end{aligned}$$

and are obviously linearly independent. Therefore, the test for incidence reduces in the worst case to five tests to find a non-zero a_j, plus six multiplications, four additions, and two zero-tests to check the two corresponding equations. If lines and points are randomly distributed in space (for most definitions of the word "random"), then on the average the incidence test will terminate after only three multiplications, two additions, and two tests for zero.

In general, if a is a flat of rank k in $\mathbf{T}_\nu$, the set of all vectors $x \in \mathbf{R}^n$ that satisfy the system of $\binom{n}{k+1}$ equations given by formula (3) is a subspace of dimension k. So, there are exactly $n-k$ linearly independent equations in that system which define $n - k$ hyperplanes whose intersection is a. Each of these hyperplanes is characterized by a label set $\mathrm{K} \subseteq \mathrm{N}$ with $k + 1$ elements; the ith coefficient of that hyperplane is

$$\begin{cases} (-1)^{|i>\mathrm{K}\backslash i|}\, a_{\mathrm{K}\backslash i} & \text{if } i \in \mathrm{K}, \\ 0 & \text{if } i \notin \mathrm{K}. \end{cases} \tag{5}$$

It turns out that if the Plücker coordinate a_J is non-zero, the hyperplanes with label sets $\mathrm{J} \cup \{i\}$, for each $i \in \mathrm{N} \backslash \mathrm{J}$, are independent. This is obvious once we realize that, among those $n - k$ hyperplanes, the one with label set $\mathrm{K} = \mathrm{J} \cup \{i\}$ is the only one with a non-zero coefficient for x_i. Notice also that each of the hyperplanes (5) has at most $k+1$ non-zero coefficients. Therefore, the point incidence test for a reduces to a search for a non-zero Plücker coordinate, followed by $(n - k)(k + 1)$ multiplications, $(n - k)k$ additions, and $n - k$ tests for zero in the worst case. For random inputs, the expected cost is a little more than $k + 1$ multiplications, k additions, and one test for zero.

1.4. Relative orientation

The formula for $a \vee b$ becomes a bit simpler when the two flats have complementary ranks, that is, when $\mathrm{rank}(a) + \mathrm{rank}(b) = n$. In that case, the result is a flat of rank n: the universe Υ of $\mathbf{T}_n$, its opposite, or (if the two flats are not disjoint) the null object $\mathbf{0}^n$.

A flat of rank n has only one Plücker coordinate c_N, where $\mathrm{N} = \{0, .. \nu\}$. Since positive scale factors do not matter, the only important property of that coordinate is its sign. We conclude that c_N is simply the relative orientation function $a \diamond b$. Let $r = \mathrm{rank}(a)$, $s = \mathrm{rank}(b) = n - r$; according to formula (2), the value of $a \diamond b$ is given by

$$c_\mathrm{N} = \sum_{\substack{\mathrm{I} \cup \mathrm{J} = \mathrm{N} \\ \mathrm{I} \cap \mathrm{J} = \emptyset \\ |\mathrm{I}| = r, |\mathrm{J}| = s}} (-1)^{|\mathrm{I} > \mathrm{J}|} a_\mathrm{I} b_\mathrm{J}$$

that is,

$$a \diamond b = \mathrm{sign}\left(\sum_{\substack{\mathrm{K} \subseteq \mathrm{N} \\ |\mathrm{K}| = r}} (-1)^{|\mathrm{K} > \overline{\mathrm{K}}|} a_\mathrm{K} b_{\overline{\mathrm{K}}} \right) \tag{6}$$

where $\overline{\mathrm{K}}$ is the set complement of K relative to $\mathrm{N} = \{0, .. \nu\}$.

1.5. Polarity

The condition for two points of $\mathbf{T}_\nu$ to be polar is obviously that the dot product of their homogeneous coordinates be zero:

$$[x_0, \ldots x_\nu] \perp [y_0, \ldots y_\nu] \Leftrightarrow x_0 y_0 + \cdots x_\nu y_\nu = 0.$$

In general, the condition for a point x to be polar to a flat a is

$$\sum_{i \in \overline{K}} (-1)^{|\{i\} > K|} x_i a_K = 0$$

for all $K \subseteq N$ with $|K| = \operatorname{rank}(a)$.

1.6. Polar complement

Let a be a flat of $\mathbf{T}_\nu$ with rank r. Its polar complements in $\mathbf{T}_\nu$ are given by

$$(a^{\vdash})_K = (-1)^{|\overline{K} > K|} a_{\overline{K}} \qquad (7)$$

$$(a^{\dashv})_K = (-1)^{|K > \overline{K}|} a_{\overline{K}} \qquad (8)$$

where $\overline{K}$ denotes the complement of K with respect to N.

1.7. Meet

Formulas for the meet of two flats can be obtained by combining those for join and polar complement. Let $r = \operatorname{rank}(a)$, $s = \operatorname{rank}(b)$, with $r + s \geq n$; from $a \wedge b = (a^{\vdash} \vee b^{\vdash})^{\dashv}$ we get

$$(a \wedge b)_K = \sum_{\substack{I \cap J = K \\ I \cup J = N \\ |I| = r, |J| = s}} (-1)^{|\overline{J} > \overline{I}|} a_I b_J \qquad (9)$$

1.8. Representative simplex

We can use the polar complement formulas above to select a representative simplex from a flat b, given its Plücker coordinates. We need only compute the polar complement $b^{\vdash}$ (formula (7)), then find a set of independent hyperplanes that contain the $b^{\vdash}$ (formula (8)), and finally list the polar complements of those hyperplanes.

2. Formulas for computers

Plücker formulas such as (9) above can be implemented as procedures that take the Plücker coordinates of the operands in natural order, as two arrays $[a_0, a_1 ..]$ and $[b_0, b_1, ..]$, and return the result C in the same format. For spaces of dimension four or less, the best policy is to write a separate routine for each combination of operand ranks, expanding the summations by hand, as shown below.

2.1. One-dimensional geometry

In one-dimensional space the join, meet, and relative orientation of two points are the same operation, but the polar complements ⊢ and ⊣ are distinct:

point ← point$^{\vdash}$
$c_0 ← -a_1$
$c_1 ←\ \ \ a_0$

line ← point $\vee$ point
vacuum ← point $\wedge$ point
sign ← point $\diamond$ point
$c_0 ← a_0 b_1 - a_1 b_0$

point ← point$^{\dashv}$
$c_0 ←\ \ \ a_1$
$c_1 ← -a_0$

2.2. Two-dimensional geometry

In two dimensions, the the join of two points and meet of two lines happen to be given by the same formulas. The relative orientation of a point and a line and that of a line and a point are also given by the same formula, and the polar complements are the same:

line ← point $\vee$ point
point ← line $\wedge$ line
$c_0 ← a_0 b_1 - a_1 b_0$
$c_1 ← a_0 b_2 - a_2 b_0$
$c_2 ← a_1 b_2 - a_2 b_1$

plane ← line $\vee$ point
plane ← point $\vee$ line
vacuum ← point $\wedge$ line
vacuum ← line $\wedge$ point
sign ← point $\diamond$ line
sign ← line $\diamond$ point
$c_0 ← a_0 b_2 - a_1 b_1 + a_2 b_0$

line ← point$^{\vdash}$
line ← point$^{\dashv}$
point ← line$^{\vdash}$
point ← line$^{\dashv}$
$c_0 ←\ \ \ a_3$
$c_1 ← -a_2$
$c_2 ←\ \ \ a_1$

2.3. Three-dimensional geometry

In three dimensions the interesting flats are points, lines, and planes. Unlike the previous cases, the join of two points and the meet of two planes are given by different formulas:

line ← point ∨ point	line ← plane ∧ plane
$c_0 \leftarrow a_0 b_1 - a_1 b_0$	$c_0 \leftarrow a_0 b_1 - a_1 b_0$
$c_1 \leftarrow a_0 b_2 - a_2 b_0$	$c_1 \leftarrow a_0 b_2 - a_2 b_0$
$c_2 \leftarrow a_1 b_2 - a_2 b_1$	$c_2 \leftarrow a_0 b_3 - a_3 b_0$
$c_3 \leftarrow a_0 b_3 - a_3 b_0$	$c_3 \leftarrow a_1 b_2 - a_2 b_1$
$c_4 \leftarrow a_1 b_3 - a_3 b_1$	$c_4 \leftarrow a_1 b_3 - a_3 b_1$
$c_5 \leftarrow a_2 b_3 - a_3 b_2$	$c_5 \leftarrow a_2 b_3 - a_3 b_2$

plane ← line ∨ point	point ← line ∧ plane
$c_0 \leftarrow a_0 b_2 - a_1 b_1 + a_2 b_0$	$c_0 \leftarrow a_0 b_2 - a_1 b_1 + a_3 b_0$
$c_1 \leftarrow a_0 b_3 - a_3 b_1 + a_4 b_0$	$c_1 \leftarrow a_0 b_3 - a_2 b_1 + a_4 b_0$
$c_2 \leftarrow a_1 b_3 - a_3 b_2 + a_5 b_0$	$c_2 \leftarrow a_1 b_3 - a_2 b_2 + a_5 b_0$
$c_3 \leftarrow a_2 b_3 - a_4 b_2 + a_5 b_1$	$c_3 \leftarrow a_3 b_3 - a_4 b_2 + a_5 b_1$

The relative orientation of a point and a plane and that of a plane and a point are also given by the same formulas, even though the operation is anticommutative. What happens is that the formula for $c \leftarrow a \vee b$ (a point, b plane) is antisymmetric in a and b. Therefore, interchanging a and b in the formula and negating everything gives back that same formula:

space ← plane ∨ point
space ← point ∨ plane
vacuum ← point ∧ plane
vacuum ← plane ∧ point
sign ← point ◇ plane
sign ← plane ◇ point
$c_0 \leftarrow a_0 b_3 - a_1 b_2 + a_2 b_1 - a_3 b_0$

Here is how we compute the relative orientation of two lines:

$$
\begin{aligned}
\text{space} &\leftarrow \text{line} \vee \text{line}\\
\text{vacuum} &\leftarrow \text{line} \wedge \text{line}\\
\text{sign} &\leftarrow \text{line} \diamond \text{line}
\end{aligned}
$$

$$
c_0 \leftarrow a_0 b_5 - a_1 b_4 + a_2 b_3 + a_3 b_2 - a_4 b_1 + a_5 b_0
$$

Since $\mathbf{T}_3$ is a space of odd dimension, the two polar complements are distinct for points and planes, but are the same for lines:

plane $\leftarrow$ point$^{\llcorner}$ point $\leftarrow$ plane$^{\llcorner}$	line $\leftarrow$ line$^{\llcorner}$ line $\leftarrow$ line$^{\lrcorner}$	plane $\leftarrow$ point$^{\lrcorner}$ point $\leftarrow$ plane$^{\lrcorner}$
$c_0 \leftarrow -a_3$	$c_0 \leftarrow a_5$	$c_0 \leftarrow a_3$
$c_1 \leftarrow a_2$	$c_1 \leftarrow -a_4$	$c_1 \leftarrow -a_2$
$c_2 \leftarrow -a_1$	$c_2 \leftarrow a_3$	$c_2 \leftarrow a_1$
$c_3 \leftarrow a_0$	$c_3 \leftarrow a_2$	$c_3 \leftarrow -a_0$
	$c_4 \leftarrow -a_1$	
	$c_5 \leftarrow a_0$	

2.4. Four-dimensional geometry

In four-dimensional geometry the complexity of the formulas begins to get prohibitive, and there are few simplifying coincidences:

line $\leftarrow$ point $\vee$ point	plane $\leftarrow$ 3-space $\wedge$ 3-space
$c_0 \leftarrow a_0 b_1 - a_1 b_0$	$c_0 \leftarrow a_0 b_1 - a_1 b_0$
$c_1 \leftarrow a_0 b_2 - a_2 b_0$	$c_1 \leftarrow a_0 b_2 - a_2 b_0$
$c_2 \leftarrow a_1 b_2 - a_2 b_1$	$c_2 \leftarrow a_0 b_3 - a_3 b_0$
$c_3 \leftarrow a_0 b_3 - a_3 b_0$	$c_3 \leftarrow a_0 b_4 - a_4 b_0$
$c_4 \leftarrow a_1 b_3 - a_3 b_1$	$c_4 \leftarrow a_1 b_2 - a_2 b_1$
$c_5 \leftarrow a_2 b_3 - a_3 b_2$	$c_5 \leftarrow a_1 b_3 - a_3 b_1$
$c_6 \leftarrow a_0 b_4 - a_4 b_0$	$c_6 \leftarrow a_1 b_4 - a_4 b_1$
$c_7 \leftarrow a_1 b_4 - a_4 b_1$	$c_7 \leftarrow a_2 b_3 - a_3 b_2$
$c_8 \leftarrow a_2 b_4 - a_4 b_2$	$c_8 \leftarrow a_2 b_4 - a_4 b_2$
$c_9 \leftarrow a_3 b_4 - a_4 b_3$	$c_9 \leftarrow a_3 b_4 - a_4 b_3$

plane ← point ∨ line	line ← 3-space ∧ plane
$c_0 \leftarrow a_0 b_2 - a_1 b_1 + a_2 b_0$	$c_0 \leftarrow a_0 b_4 - a_1 b_1 + a_2 b_0$
$c_1 \leftarrow a_0 b_4 - a_1 b_3 + a_3 b_0$	$c_1 \leftarrow a_0 b_5 - a_1 b_2 + a_3 b_0$
$c_2 \leftarrow a_0 b_5 - a_2 b_3 + a_3 b_1$	$c_2 \leftarrow a_0 b_6 - a_1 b_3 + a_4 b_0$
$c_3 \leftarrow a_1 b_5 - a_2 b_4 + a_3 b_2$	$c_3 \leftarrow a_0 b_7 - a_2 b_2 + a_3 b_1$
$c_4 \leftarrow a_0 b_7 - a_1 b_6 + a_4 b_0$	$c_4 \leftarrow a_0 b_8 - a_2 b_3 + a_4 b_1$
$c_5 \leftarrow a_0 b_8 - a_2 b_6 + a_4 b_1$	$c_5 \leftarrow a_0 b_9 - a_3 b_3 + a_4 b_2$
$c_6 \leftarrow a_1 b_8 - a_2 b_7 + a_4 b_2$	$c_6 \leftarrow a_1 b_7 - a_2 b_5 + a_3 b_4$
$c_7 \leftarrow a_0 b_9 - a_3 b_6 + a_4 b_3$	$c_7 \leftarrow a_1 b_8 - a_2 b_6 + a_4 b_4$
$c_8 \leftarrow a_1 b_9 - a_3 b_7 + a_4 b_4$	$c_8 \leftarrow a_1 b_9 - a_3 b_6 + a_4 b_5$
$c_9 \leftarrow a_2 b_9 - a_3 b_8 + a_4 b_5$	$c_9 \leftarrow a_2 b_9 - a_3 b_8 + a_4 b_7$

3-space ← point ∨ plane	point ← 3-space ∧ line
$c_0 \leftarrow a_0 b_3 - a_1 b_2 + a_2 b_1 - a_3 b_0$	$c_0 \leftarrow a_0 b_6 - a_1 b_3 + a_2 b_1 - a_3 b_0$
$c_1 \leftarrow a_0 b_6 - a_1 b_5 + a_2 b_4 - a_4 b_0$	$c_1 \leftarrow a_0 b_7 - a_1 b_4 + a_2 b_2 - a_4 b_0$
$c_2 \leftarrow a_0 b_8 - a_1 b_7 + a_3 b_4 - a_4 b_1$	$c_2 \leftarrow a_0 b_8 - a_1 b_5 + a_3 b_2 - a_4 b_1$
$c_3 \leftarrow a_0 b_9 - a_2 b_7 + a_3 b_5 - a_4 b_2$	$c_3 \leftarrow a_0 b_9 - a_2 b_5 + a_3 b_4 - a_4 b_3$
$c_4 \leftarrow a_1 b_9 - a_2 b_8 + a_3 b_6 - a_4 b_3$	$c_4 \leftarrow a_1 b_9 - a_2 b_8 + a_3 b_7 - a_4 b_6$

3-space ← plane ∨ point	point ← line ∧ 3-space
$c_0 \leftarrow a_0 b_3 - a_1 b_2 + a_2 b_1 - a_3 b_0$	$c_0 \leftarrow a_0 b_3 - a_1 b_2 + a_3 b_1 - a_6 b_0$
$c_1 \leftarrow a_0 b_4 - a_4 b_2 + a_5 b_1 - a_6 b_0$	$c_1 \leftarrow a_0 b_4 - a_2 b_2 + a_4 b_1 - a_7 b_0$
$c_2 \leftarrow a_1 b_4 - a_4 b_3 + a_7 b_1 - a_8 b_0$	$c_2 \leftarrow a_1 b_4 - a_2 b_3 + a_5 b_1 - a_8 b_0$
$c_3 \leftarrow a_2 b_4 - a_5 b_3 + a_7 b_2 - a_9 b_0$	$c_3 \leftarrow a_3 b_4 - a_4 b_3 + a_5 b_2 - a_9 b_0$
$c_4 \leftarrow a_3 b_4 - a_6 b_3 + a_8 b_2 - a_9 b_1$	$c_4 \leftarrow a_6 b_4 - a_7 b_3 + a_8 b_2 - a_9 b_1$

plane ← line ∨ point	line ← plane ∧ 3-space
$c_0 \leftarrow a_0 b_2 - a_1 b_1 + a_2 b_0$	$c_0 \leftarrow a_0 b_2 - a_1 b_1 + a_4 b_0$
$c_1 \leftarrow a_0 b_3 - a_3 b_1 + a_4 b_0$	$c_1 \leftarrow a_0 b_3 - a_2 b_1 + a_5 b_0$
$c_2 \leftarrow a_1 b_3 - a_3 b_2 + a_5 b_0$	$c_2 \leftarrow a_0 b_4 - a_3 b_1 + a_6 b_0$
$c_3 \leftarrow a_2 b_3 - a_4 b_2 + a_5 b_1$	$c_3 \leftarrow a_1 b_3 - a_2 b_2 + a_7 b_0$
$c_4 \leftarrow a_0 b_4 - a_6 b_1 + a_7 b_0$	$c_4 \leftarrow a_1 b_4 - a_3 b_2 + a_8 b_0$
$c_5 \leftarrow a_1 b_4 - a_6 b_2 + a_8 b_0$	$c_5 \leftarrow a_2 b_4 - a_3 b_3 + a_9 b_0$
$c_6 \leftarrow a_2 b_4 - a_7 b_2 + a_8 b_1$	$c_6 \leftarrow a_4 b_3 - a_5 b_2 + a_7 b_1$
$c_7 \leftarrow a_3 b_4 - a_6 b_3 + a_9 b_0$	$c_7 \leftarrow a_4 b_4 - a_6 b_2 + a_8 b_1$
$c_8 \leftarrow a_4 b_4 - a_7 b_3 + a_9 b_1$	$c_8 \leftarrow a_5 b_4 - a_6 b_3 + a_9 b_1$
$c_9 \leftarrow a_5 b_4 - a_8 b_3 + a_9 b_2$	$c_9 \leftarrow a_7 b_4 - a_8 b_3 + a_9 b_2$

3-space ← line ∨ line
$c_0 \leftarrow a_0 b_5 - a_1 b_4 + a_2 b_3 + a_3 b_2 - a_4 b_1 + a_5 b_0$
$c_1 \leftarrow a_0 b_8 - a_1 b_7 + a_2 b_6 + a_6 b_2 - a_7 b_1 + a_8 b_0$
$c_2 \leftarrow a_0 b_9 - a_3 b_7 + a_4 b_6 + a_6 b_4 - a_7 b_3 + a_9 b_0$
$c_3 \leftarrow a_1 b_9 - a_3 b_8 + a_5 b_6 + a_6 b_5 - a_8 b_3 + a_9 b_1$
$c_4 \leftarrow a_2 b_9 - a_4 b_8 + a_5 b_7 + a_7 b_5 - a_8 b_4 + a_9 b_2$

point ← plane ∧ plane
$c_0 \leftarrow a_0 b_7 - a_1 b_5 + a_2 b_4 + a_4 b_2 - a_5 b_1 + a_7 b_0$
$c_1 \leftarrow a_0 b_8 - a_1 b_6 + a_3 b_4 + a_4 b_3 - a_6 b_1 + a_8 b_0$
$c_2 \leftarrow a_0 b_9 - a_2 b_6 + a_3 b_5 + a_5 b_3 - a_6 b_2 + a_9 b_0$
$c_3 \leftarrow a_1 b_9 - a_2 b_8 + a_3 b_7 + a_7 b_3 - a_8 b_2 + a_9 b_1$
$c_4 \leftarrow a_4 b_9 - a_5 b_8 + a_6 b_7 + a_7 b_6 - a_8 b_5 + a_9 b_4$

The relative orientations of points vs. 3-space and 3-space vs. point are given by the same formula. Again, it is a case of both the formula and the operation being antisymmetric in a and b:

4-space ← 3-space ∨ point
4-space ← point ∨ 3-space
vacuum ← point ∧ 3-space
vacuum ← 3-space ∧ point
sign ← point ◇ 3-space
sign ← 3-space ◇ point
$c_0 \leftarrow a_0 b_4 - a_1 b_3 + a_2 b_2 - a_3 b_1 + a_4 b_0$

On the other hand, the formulas for line vs. plane and plane vs. line are different, even though the operation itself is commutative:

4-space ← line ∨ plane
vacuum ← line ∧ plane
sign ← line ◇ plane
$c_0 \leftarrow a_0 b_9 - a_1 b_8 + a_2 b_7 + a_3 b_6 - a_4 b_5 + a_5 b_4 - a_6 b_3 + a_7 b_2 - a_8 b_1 + a_9 b_0$

$$4\text{-space} \leftarrow \text{plane} \vee \text{line}$$
$$\text{vacuum} \leftarrow \text{plane} \wedge \text{line}$$
$$\text{sign} \leftarrow \text{plane} \diamond \text{line}$$

$$c_0 \leftarrow a_0 b_9 - a_1 b_8 + a_2 b_7 - a_3 b_6 + a_4 b_5 - a_5 b_4 + a_6 b_3 + a_7 b_2 - a_8 b_1 + a_9 b_0$$

The two polar complements coincide:

plane $\leftarrow$ line$^\vdash$		line $\leftarrow$ plane$^\vdash$
plane $\leftarrow$ line$^\dashv$	3-space $\leftarrow$ point$^\vdash$	line $\leftarrow$ plane$^\dashv$
$c_0 \leftarrow a_9$	3-space $\leftarrow$ point$^\dashv$	$c_0 \leftarrow a_9$
$c_1 \leftarrow -a_8$	point $\leftarrow$ 3-space$^\vdash$	$c_1 \leftarrow -a_8$
$c_2 \leftarrow a_7$	point $\leftarrow$ 3-space$^\dashv$	$c_2 \leftarrow a_7$
$c_3 \leftarrow -a_6$		$c_3 \leftarrow a_6$
$c_4 \leftarrow a_5$	$c_0 \leftarrow a_4$	$c_4 \leftarrow -a_5$
$c_5 \leftarrow -a_4$	$c_1 \leftarrow -a_3$	$c_5 \leftarrow a_4$
$c_6 \leftarrow a_3$	$c_2 \leftarrow a_2$	$c_6 \leftarrow -a_3$
$c_7 \leftarrow a_2$	$c_3 \leftarrow -a_1$	$c_7 \leftarrow a_2$
$c_8 \leftarrow -a_1$	$c_4 \leftarrow a_0$	$c_8 \leftarrow -a_1$
$c_9 \leftarrow a_0$		$c_9 \leftarrow a_0$

Note that the formulas for join in $(\nu - 1)$ dimensions can be obtained from those join in ν dimensions by dropping all terms that involve non-existing coordinates.

3. Projective maps in Plücker coordinates

As we saw in chapter 8, computing the image of a point by a projective map means post-multiplying its homogeneous coordinates by the associate matrix. The following theorem generalizes this result to flats of arbitrary rank:

Theorem 1. *The image of a flat* $[a_0, a_1 \ldots]^k$ *by a projective map* $M = \llbracket M \rrbracket$ *is the flat* $[b_0, b_1, \ldots]^k$ *where*

$$b_j = \sum_i a_i (M^{(k)})_j^i \tag{10}$$

and $M^{(k)}$ *is the* $\binom{n}{k} \times \binom{n}{k}$ *matrix whose elements are all* $k \times k$ *minor determinants of* M, *in natural order of row and column labels.*

PROOF: Consider a κ-dimensional flat a with coordinates $[a_0, a_1, \ldots]^k$. Let $u = (u^0; \ldots u^\kappa)$ be a representative simplex of a, with $u^i = [u_0^i, \ldots u_\nu^i]$, viewed as a

$k \times n$ matrix. Recall that a_i is the coordinate whose label is the integer set $I = k{:}i$, that is, the determinant of the $k \times k$ minor formed by taking the columns of u whose indices are in the set I.

Let $aM = b = [b_0, .. b_{m-1}]^k$. A representative simplex of b is $v = (v^0; .. v^\kappa)$ where $v_j^i = \sum_t u_t^i M_j^t$. Therefore, the Plücker coordinates of b are given by the determinants

$$
b_J =
\begin{vmatrix}
v_{j_0}^0 & v_{j_1}^0 & \cdots & v_{j_\kappa}^0 \\
v_{j_0}^1 & v_{j_1}^1 & \cdots & v_{j_\kappa}^1 \\
\vdots & & & \vdots \\
v_{j_0}^\kappa & v_{j_1}^\kappa & \cdots & v_{j_\kappa}^\kappa
\end{vmatrix}
$$

$$
=
\begin{vmatrix}
\sum_{i_0} u_{i_0}^0 M_{j_0}^{i_0} & \sum_{i_1} u_{i_1}^0 M_{j_1}^{i_1} & \cdots & \sum_{i_\kappa} u_{i_\kappa}^0 M_{j_\kappa}^{i_\kappa} \\
\sum_{i_0} u_{i_0}^1 M_{j_0}^{i_0} & \sum_{i_1} u_{i_1}^1 M_{j_1}^{i_1} & \cdots & \sum_{i_\kappa} u_{i_\kappa}^1 M_{j_\kappa}^{i_\kappa} \\
\vdots & & & \vdots \\
\sum_{i_0} u_{i_0}^\kappa M_{j_0}^{i_0} & \sum_{i_1} u_{i_1}^\kappa M_{j_1}^{i_1} & \cdots & \sum_{i_\kappa} u_{i_\kappa}^\kappa M_{j_\kappa}^{i_\kappa}
\end{vmatrix}
\tag{11}
$$

where J ranges over all k-element subsets of N, $j_0, j_1, .. j_\kappa$ are the elements of J in increasing order, and each i_k ranges from 0 to ν. Note that each column of (11) is a linear combination of columns of the u matrix. Since the determinant of a matrix is a multilinear function of its columns, we can expand equation (11) into

$$
b_J = \sum_{i_0} \sum_{i_1} \cdots \sum_{i_\kappa}
\begin{vmatrix}
u_{i_0}^0 & u_{i_1}^0 & \cdots & u_{i_\kappa}^0 \\
u_{i_0}^1 & u_{i_1}^1 & \cdots & u_{i_\kappa}^1 \\
\vdots & & & \vdots \\
u_{i_0}^\kappa & u_{i_1}^\kappa & \cdots & u_{i_\kappa}^\kappa
\end{vmatrix}
\cdot M_{j_0}^{i_0} M_{j_1}^{i_1} \cdots M_{j_\kappa}^{i_\kappa}
\tag{12}
$$

where the indices i_k still range independently over $0 .. \nu$.

Every term of summation (12) where two indices i_p, i_q are equal is zero. Moreover, two sequences $i_0 \ldots i_\kappa$ which differ only on the order of the elements will give rise to two terms that differ at most in their signs. More precisely, formula (12) is equivalent to

$$
b_{\mathrm{J}} = \sum_{0 \le i_0 < i_1 < \cdots < i_\kappa \le \nu}
\begin{vmatrix}
u^0_{i_0} & u^0_{i_1} & \cdots & u^0_{i_\kappa} \\
u^1_{i_0} & u^1_{i_1} & \cdots & u^1_{i_\kappa} \\
\vdots & & & \vdots \\
u^\kappa_{i_0} & u^\kappa_{i_1} & \cdots & u^\kappa_{i_\kappa}
\end{vmatrix}
\cdot \sum_\pi (-1)^{\|\pi\|} \cdot M^{i_{\pi(0)}}_{j_0} M^{i_{\pi(1)}}_{j_1} \cdots M^{i_{\pi(\kappa)}}_{j_\kappa}
$$

where π ranges over all permutations of $0..\kappa$, and $\|\pi\|$ is the number of inversions in π. But the second summation is simply the minor determinant formed by lines $j_0 \ldots j_\kappa$ and columns $i_0 \ldots i_\kappa$ of matrix M:

$$
b_{\mathrm{J}} = \sum_{0 \le i_0 < i_1 < \cdots < i_\kappa \le \nu}
\begin{vmatrix}
u^0_{i_0} & u^0_{i_1} & \cdots & u^0_{i_\kappa} \\
u^1_{i_0} & u^1_{i_1} & \cdots & u^1_{i_\kappa} \\
\vdots & & \vdots \\
u^\kappa_{i_0} & u^\kappa_{i_1} & \cdots & u^\kappa_{i_\kappa}
\end{vmatrix}
\cdot
\begin{vmatrix}
M^{i_0}_{j_0} & M^{i_1}_{j_0} & \cdots & M^{i_\kappa}_{j_0} \\
M^{i_0}_{j_1} & M^{i_1}_{j_1} & \cdots & M^{i_\kappa}_{j_1} \\
\vdots & & & \vdots \\
M^{i_0}_{j_\kappa} & M^{i_1}_{j_\kappa} & \cdots & M^{i_\kappa}_{j_\kappa}
\end{vmatrix}
$$

Therefore, we conclude that

$$
b_{\mathrm{J}} = \sum_{\mathrm{I}} a_{\mathrm{I}} (M^{(k)})^{\mathrm{I}}_{\mathrm{J}}
$$

where the index set I ranges over all k-element subsets of $\{0 .. \nu\}$.
QED.

For example, if M is the matrix

$$
\begin{pmatrix}
1 & 3 & 5 & 1 \\
2 & 1 & 0 & 2 \\
1 & 2 & 0 & 4 \\
1 & 0 & 0 & 1
\end{pmatrix}
$$

then $M^{(2)}$ is the matrix

$$
M^{(2)} =
\begin{pmatrix}
\begin{vmatrix}1&3\\2&1\end{vmatrix} & \begin{vmatrix}1&5\\2&0\end{vmatrix} & \begin{vmatrix}3&5\\1&0\end{vmatrix} & \begin{vmatrix}1&1\\2&2\end{vmatrix} & \begin{vmatrix}3&1\\1&2\end{vmatrix} & \begin{vmatrix}5&1\\0&2\end{vmatrix} \\[6pt]
\begin{vmatrix}1&3\\1&2\end{vmatrix} & \begin{vmatrix}1&5\\1&0\end{vmatrix} & \begin{vmatrix}3&5\\2&0\end{vmatrix} & \begin{vmatrix}1&1\\1&4\end{vmatrix} & \begin{vmatrix}3&1\\2&4\end{vmatrix} & \begin{vmatrix}5&1\\0&4\end{vmatrix} \\[6pt]
\begin{vmatrix}2&1\\1&2\end{vmatrix} & \begin{vmatrix}2&0\\1&0\end{vmatrix} & \begin{vmatrix}1&0\\2&0\end{vmatrix} & \begin{vmatrix}2&2\\1&4\end{vmatrix} & \begin{vmatrix}1&2\\2&4\end{vmatrix} & \begin{vmatrix}0&2\\0&4\end{vmatrix} \\[6pt]
\begin{vmatrix}1&3\\1&0\end{vmatrix} & \begin{vmatrix}1&5\\1&0\end{vmatrix} & \begin{vmatrix}3&5\\0&0\end{vmatrix} & \begin{vmatrix}1&1\\1&1\end{vmatrix} & \begin{vmatrix}3&1\\0&1\end{vmatrix} & \begin{vmatrix}5&1\\0&1\end{vmatrix} \\[6pt]
\begin{vmatrix}2&1\\1&0\end{vmatrix} & \begin{vmatrix}2&0\\1&0\end{vmatrix} & \begin{vmatrix}1&0\\0&0\end{vmatrix} & \begin{vmatrix}2&2\\1&1\end{vmatrix} & \begin{vmatrix}1&2\\0&1\end{vmatrix} & \begin{vmatrix}0&2\\0&1\end{vmatrix} \\[6pt]
\begin{vmatrix}1&2\\1&0\end{vmatrix} & \begin{vmatrix}1&0\\1&0\end{vmatrix} & \begin{vmatrix}2&0\\0&0\end{vmatrix} & \begin{vmatrix}1&4\\1&1\end{vmatrix} & \begin{vmatrix}2&4\\0&1\end{vmatrix} & \begin{vmatrix}0&4\\0&1\end{vmatrix}
\end{pmatrix}
$$

$$
=
\begin{pmatrix}
-5 & -10 & -5 & 0 & 5 & 10 \\
-1 & -5 & -10 & 3 & 10 & 20 \\
3 & 0 & 0 & 6 & 0 & 0 \\
-3 & -5 & 0 & 0 & 3 & 5 \\
2 & 0 & 0 & 0 & 1 & 0 \\
-2 & 0 & 0 & -3 & 2 & 0
\end{pmatrix}
$$

This result is of little practical value, since computing the matrix $M^{(k)}$ and applying it is much too expensive. In general it is far more efficient to use the simplex representation (reduced or not), and map each vertex of the simplex through the original matrix M. This takes $O(kn^2)$ time, including the $O(k^2 n)$ cost of putting the mapped simplex in reduced form. Even if the given flat and the answer are represented by Plücker coordinates, it is more efficient in general to extract from them a representative simplex, apply the map to each vertex of the latter, and compute the Plücker coordinates of the result by the join formula.

This is already true even in the simplest non-trivial case, namely lines in three-space ($n = 3, k = 2$). Computing the matrix $M^{(2)}$ costs 72 multiplications (μ) and 36 additions/subtractions (α). Once we have this matrix, the cost of mapping a line is $36\mu + 30\alpha$. If instead we compute two independent points from the Plücker coordinates, map those points through M, and compute the join of the images, the total cost will be about $8(3\mu+2\alpha)+6(\mu+2\alpha) = 30\mu+28\alpha$, plus some logic overhead. (Note that the representative points computed as described in section 1.8 have only three non-zero coordinates). By comparison, mapping an arbitrary pair of points (an unreduced simplex) through the map M costs $32\mu + 24\alpha$. Mapping a reduced simplex and reducing the result costs in general $4(2\mu+2\alpha)+10\mu+5\alpha = 18\mu+13\alpha$. In spaces of higher dimensions Plücker coordinates are even more expensive, because the size of the matrix $M^{(k)}$, $\binom{n}{k} \times \binom{n}{k}$, grows exponentially with n and k.

4. Directions and parallelism

The direction of a flat is easily computed from its Plücker coordinates. Recall that the coordinates of Ω are $[0,..0,1]^{n-1}$. Therefore, the direction of a flat a of rank r is the flat $\operatorname{dir}(a)$ of rank $r-1$ whose coordinates are

$$(\operatorname{dir} a)_K = (a \wedge \Omega)_K$$

$$= \sum_{\substack{I\cap J=K \\ I\cup J=\{0..\nu\} \\ |I|=r \\ |J|=n-1}} (-1)^{|\bar J>\bar I|} a_I \, \Omega_J$$

$$= \sum_{\substack{I\cap\{1..\nu\}=K \\ I\cup\{1..\nu\}=\{0..\nu\} \\ |I|=r}} (-1)^{|\{0\}>\bar I|} a_I$$

$$= \begin{cases} 0 & \text{if } 0 \in K \\ a_{\{0\}\cup K} & \text{if } 0 \notin K \end{cases} \tag{13}$$

For example, the direction of a line l in three-space is the point x with coordinates

$$x_{\{0\}} = 0, \quad x_{\{1\}} = l_{\{01\}}, \quad x_{\{2\}} = l_{\{02\}}, \quad x_{\{3\}} = l_{\{03\}}$$

or, in positional notation, $\operatorname{dir}[l_0,..l_5]^2 = [0, l_0, l_1, l_3]$. In the same way we get formulas for the direction of a line of $\mathbf{T}_2$ or a plane of $\mathbf{T}_3$:

$$\operatorname{dir}[l_0, l_1, l_2]^2 = [0, l_0, l_1]^1$$
$$\operatorname{dir}[h_0, h_1, h_2, h_3]^3 = [0, 0, h_0, 0, h_1, h_2]^2$$

4.1. Parallelism

Recall that $a \mathbin{\text{⇈}} b$ was defined as a shorthand for $\operatorname{dir} a = \operatorname{dir} b$. From formula (13), we see that $a \mathbin{\text{⇈}} b$ if and only if the Plücker coordinates of a whose label set includes 0 are a positive multiple of the corresponding coordinates of b. For example, if a and b are two lines in three-space, we have $[a_0, .. a_5]^2 \mathbin{\text{⇈}} [b_0, .. b_5]^2$ if and only if

$$a_0 = \lambda b_0 \quad \text{and} \quad a_1 = \lambda b_1 \quad \text{and} \quad a_3 = \lambda b_3$$

for some $\lambda > 0$. If f is a proper flat of $\mathbf{T}_\nu$ with rank r, and p is a point on the front range, then the flat passing through p and with same rank and direction as f has coordinates given by the formula

$$(p \vee \operatorname{dir} f)_K = \sum_{\substack{I \cup J = K \\ I \cap J = \varnothing \\ |I| = 1, |J| = r-1}} (-1)^{|I > J|} p_I (\operatorname{dir} f)_J$$

$$= \sum_{i \in K} (-1)^{|i > K \setminus i|} p_i (\operatorname{dir} f)_{K \setminus i}$$

$$= \sum_{i \in K} (-1)^{|i > K \setminus i|} p_i \cdot \begin{cases} 0 & \text{if } 0 \in K \setminus i \\ f_{\{0\} \cup K \setminus i} & \text{if } 0 \notin K \setminus i \end{cases}$$

$$= \begin{cases} p_0 f_K & \text{if } 0 \in K \\ \sum_{i \in K} (-1)^{|i > K \setminus i|} p_i f_{\{0\} \cup K \setminus i} & \text{if } 0 \notin K \end{cases}$$

For example, in three-space the line through $p = [p_0, .. p_3]$ and parallel to the line $l = [l_0, .. l_5]^2$ is

$$[p_0 l_0,\ p_0 l_1,\ p_1 l_1 - p_2 l_0,\ p_0 l_3,\ p_1 l_3 - p_3 l_0,\ p_2 l_3 - p_3 l_1]^2 \tag{14}$$

The analogous formula for two-dimensional geometry is obtained by dropping the last three coordinates, that is,

$$[p_0 l_0,\ p_0 l_1,\ p_1 l_1 - p_2 l_0]^2$$

The plane through p parallel to $h = [h_0, .. h_3]^3$ is

$$[p_0 h_0,\ p_1 h_1,\ p_2 h_2,\ p_1 h_2 - p_2 h_1 + p_3 h_0]^3$$

References

[1] C. F. Adler: **Modern geometry** (2nd. ed.). McGraw-Hill (1970).

[2] M. Barnabei, A. Brini, and G.-C. Rota: **On the exterior calculus of invariant theory.** Journal of Algebra vol. 96 no. 1 (Sept. 1985), 120–160.

[3] G. Berman: **The wedge product.** American Mathematical Monthly vol. 68 no. 2 (1961), 112–119.

[4] L. M. Blumenthal and K. Menger: **Studies in geometry.** W. H. Freeman (1970).

[5] K. Q. Brown: **Voronoi diagrams from convex hulls.** Information Processing Letters vol. 9 no. 5 (December 1979), 223–228.

[6] H. S. M. Coxeter: **The real projective plane.** McGraw-Hill (1949).

[7] H. S. M. Coxeter: **Regular Polytopes (2nd edition).** Dover, New York (1973).

[8] H. Crapo and J. Ryan: **Scene analysis and geometric homology.** Proceedings of the 2nd ACM Symposium on Computational Geometry (June 1986), 125–132.

[9] H. Edelsbrunner: **Algorithms in combinatorial geometry.** Springer-Verlag, Berlin (1987).

[10] G. H. Golub and C. F. Van Loan: **Matrix computations.** Johns Hopkins University Press, Baltimore (1983).

[11] B. Grünbaum: **Convex polytopes.** John Wiley & Sons, London (1967).

[12] P. Hanrahan: **A homogeneous geometry calculator.** Technical memo 3-D no. 7, Computer Graphics Laboratory, New York Institute of Technology (September 1984).

[13] D. Hestenes and G. Sobczyk: **Clifford algebra to geometric calculus: A unified language for mathematics and physics.** D. Reidel, Dordrecht (1984).

[14] W. V. D. Hodge, D. Pedoe: **Methods of algebraic geometry.** Cambridge University Press (1952).

[15] P. S. Modenov and A. S. Parkhomenko: **Geometric transformations. vol. II: Projective transformations.** Academic Press (1965).

[16] A. Penna and R. Patterson: **Projective geometry and its applications to computer graphics.** Prentice-Hall (1986).

[17] F. P. Preparata and D. E. Muller: **Finding the intersection of n half-spaces in time $O(n \log n)$.** Theoretical Computer Science vol. 8 (1979), 44–55

[18] W. H. Press, B. P. Flannery, S. A. Teukolsky, and W. T. Vetterling: **Numerical recipes: The art of scientific computing.** Cambridge University Press, Cambridge (1986).

[19] R. F. Riesenfeld: **Homogeneous coordinates and projective planes in computer graphics.** IEEE Computer Graphics and Automation (Jan. 1981), 50–55.

[20] I. M. Yaglom and V. G. Boltyanskii: **Convex figures.** Holt, Rinehart and Winston, New York (1961).

List of symbols

Symbol	Meaning	Page
$a^{\vdash}$	right polar complement of flat a relative to Υ	85
$a^{\dashv}$	left polar complement of flat a relative to Υ	85
$a \rceil f$	right polar complement of flat a relative to flat f	91
$f \lceil a$	left polar complement of flat a relative to flat f	91
$Null(M)$	null space of map M	95
$Range(M)$	range of map M	95
$Dom(M)$	natural domain of map M	98
$Span(a)$	flat set spanned by arrangement a	109
pfr_σ, mfr_σ	standard point/mixed frame with signature σ	114
$Hull(X)$	convex hull of set X	141
$X \uplus Y$	convex hull of $X \cup Y$	142
$X^{\triangleleft}, X^{\triangleright}$	support of points, or kernel of hyperplanes	144
X^*	convex dual of convex set X	144
$\mathbf{A}_\nu$	ν-dimensional two-sided affine space	151
afr_σ	standard affine frame with signature σ	114
$dir(a)$	direction of flat a	153
$a \parallel b$	flats a and b are parallel	154
$a \Uparrow b$, $a \Updownarrow b$	flats a and b have same/opposite directions	154
$vol(s)$	measure of simplex s	164
$\mathbf{V}_\nu$	ν-dimensional two-sided vector space	157
O_ν	point $[1, 0, \ldots, 0]$, the front origin of $\mathbf{T}_\nu$	167
$norm(a)$	normal direction of flat a	174
$len(v)$	length of two-sided vector v	183
$dist(x, y)$	distance between points x and y	184
$cls(x, y)$	closeness between x and y ($= 1/dist(x, y)$)	185
$shr(v)$	shortness of vector v ($= 1/len(v)$)	185
$arg(d)$	angle between $\mathbf{e}^1$ and the direction d	187
$u \mathbin{\hat{+}} v$	sum of angles u and v	188
$u \mathbin{\hat{-}} v$	difference of angles u and v	188
$binv(X)$	binary value of set $s \subseteq \mathbf{N}$	200
$k{:}i$	ith subset of $\mathbf{N}$ with k elements	200
$\#X$	index of X among the k-element subsets of $\mathbf{N}$	200
$[z_0, z_1, \ldots]^k$	flat of rank k with Plücker coordinates $z_0, z_1, \ldots$	200
$\mathbf{e}^X$	canonical flat with label set X	202
$\langle z^0, z^1, \ldots \rangle^k$	flat of rank k with Plücker coefficients $z_0, z_1, \ldots$	203

Index